ISBN 0-8373-6662-3

DANTES– 62

Rudman's Question and Answer

DANTES
SUBJECT STANDARDIZED TESTS

Subject Examination In . . .

GEOGRAPHY

Questions and Answers

NLC

NATIONAL LEARNING CORPORATION
212 MICHAEL DRIVE, SYOSSET, NEW YORK 11791 (516) 921-8888

[redacted]

National Learning Corporation

212 Michael Drive, Syosset, New York 11791

(516) 921-8888
Outside N.Y.: 1(800) 645-6337
ORDER FAX: 1(516) 921-8743
www.passbooks.com
email: passbooks @ aol.com
sales @ passbooks.com
info @ passbooks.com

PRINTED IN THE UNITED STATES OF AMERICA

PASSBOOK®

NOTICE

This book is *SOLELY* intended for, is sold *ONLY* to, and its use is *RESTRICTED* to *individual*, bona fide applicants or candidates who qualify by virtue of having seriously filed applications for appropriate license, certificate, professional and/or promotional advancement, higher school matriculation, scholarship, or other legitimate requirements of educational and/or governmental authorities.

This book is *NOT* intended for use, class instruction, tutoring, training, duplication, copying, reprinting, excerption, or adaptation, etc., by:

(1) Other Publishers
(2) Proprietors and/or Instructors of "Coaching" and/or Preparatory Courses
(3) Personnel and/or Training Divisions of commercial, industrial, and governmental organizations
(4) Schools, colleges, or universities and/or their departments and staffs, including teachers and other personnel
(5) Testing Agencies or Bureaus
(6) Study groups which seek by the purchase of a single volume to copy and/or duplicate and/or adapt this material for use by the group as a whole without having purchased individual volumes for each of the members of the group
(7) Et al.

Such persons would be in violation of appropriate Federal and State statutes.

PROVISION OF LICENSING AGREEMENTS. — Recognized educational commercial, industrial, and governmental institutions and organizations, and others legitimately engaged in educational pursuits, including training, testing, and measurement activities, may address a request for a licensing agreement to the copyright owners, who will determine whether, and under what conditions, including fees and charges, the materials in this book may be used by them. In other words, a licensing facility *exists* for the legitimate use of the material in this book on other than an individual basis. However, it is asseverated and affirmed here that the materials in this book *CANNOT* be used without the receipt of the express permission of such a licensing agreement from the Publishers.

NATIONAL LEARNING CORPORATION
212 Michael Drive
Syosset, New York 11791

Inquiries re licensing agreements should be addressed to:
The President
National Learning Corporation
212 Michael Drive
Syosset, New York 11791

PASSBOOK SERIES®

THE *PASSBOOK SERIES®* has been created to prepare applicants and candidates for the ultimate academic battlefield—the examination room.

At some time in our lives, each and every one of us may be required to take an examination—for validation, matriculation, admission, qualification, registration, certification, or licensure.

Based on the assumption that every applicant or candidate has met the basic formal educational standards, has taken the required number of courses, and read the necessary texts, the *PASSBOOK SERIES®* furnishes the one special preparation which may assure passing with confidence, instead of failing with insecurity. Examination questions—together with answers—are furnished as the basic vehicle for study so that the mysteries of the examination and its compounding difficulties may be eliminated or diminished by a sure method.

This book is meant to help you pass your examination provided that you qualify and are serious in your objective.

The entire field is reviewed through the huge store of content information which is succinctly presented through a provocative and challenging approach—the question-and-answer method.

A climate of success is established by furnishing the correct answers at the end of each test.

You soon learn to recognize types of questions, forms of questions, and patterns of questioning. You may even begin to anticipate expected outcomes.

You perceive that many questions are repeated or adapted so that you gain acute insights, which may enable you to score many sure points.

You learn how to confront new questions, or types of questions, and to attack them confidently and work out the correct answers.

You note objectives and emphases, and recognize pitfalls and dangers, so that you may make positive educational adjustments.

Moreover, you are kept fully informed in relation to new concepts, methods, practices, and directions in the field.

You discover that you are actually taking the examination all the time: you are preparing for the examination by "taking" an examination, not by reading extraneous and/or supererogatory textbooks.

In short, this PASSBOOK®, used directedly, should be an important factor in helping you to pass your test.

DANTES Subject Standardized Tests

INTRODUCTION

The DANTES (Defense Activity for Non-Traditional Education Support) subject standardized tests are comprehensive college and graduate level examinations given by the Armed Forces, colleges and graduate schools as end-of-subject course evaluation final examinations or to obtain college equivalency credits in the various subject areas tested.

Active military personnel may take CLEP (College-Level Examination Program) examinations without charge through the DANTES program. Information about DANTES and CLEP may be obtained from the Test Control Officers at military installations, and from Educational Testing Service in Princeton, New Jersey.

The DANTES Examination Program enables students to obtain college credit for what they have learned on the job, through self-study, personal interest, correspondence courses or by any other means. It is used by colleges and universities to award college credit to students who demonstrate that they know as much as students completing an equivalent college course. It is a cost-efficient, time-saving way for students to use their knowledge to accomplish their educational goals.

Most schools accept the American Council on Education (ACE) recommendations for the minimum score required and the amount of credit awarded, but not all schools do. Be sure to check the policy regarding the score level required for credit and the number of credits to be awarded.

Not all tests are accepted by all institutions. Even when a test is accepted by an institution, it may not be acceptable for every program at that institution. Before considering testing, ascertain the acceptability of a specific test for a particular course.

Colleges and universities that administer DANTES tests may administer them to any applicant -- or they may administer the tests only to students registered at their institution. Decisions about who will be allowed to test are made by the school. Students should contact the test center to determine current policies and schedules for DANTES testing.

Colleges and universities authorized to administer DANTES tests usually do so throughout the calendar year. Each school sets its own fee for test administration and establishes its own testing schedule. Contact the representative at the administering school directly to make arrangements for testing.

Your Experience Can Earn You Credit.

It's a fact. People learn in many ways . . . through hobbies . . . interests . . . reading . . . work. The DANTES Program is a way for you to "cash in" on what you've learned from *your* experiences.

Take advantage of DANTES tests to earn college credit for what you know. More than 30 subjects are offered. There are bound to be areas where your knowledge can pay off academically.

Here are just a few of the program's benefits for you:

- ***Time and Money Saver***
  You don't have to sit through classes and pay for courses when you already know most of what's being taught and can learn the rest yourself.
- ***Credits Widely Accepted***
  Hundreds of colleges nationwide now give credit for DANTES tests. And the list is growing!
- ***Quicker Route to the Degree***
  Your dreams are a little closer than you thought possible. With DANTES.
- ***Start Fast***
  No need to wait to take advanced courses. Passing DANTES tests gives you a real head start.

Sound Good? Here's How to Get Going!

Check with your college to see if it administers the DANTES program. If the college doesn't but would like to, ask that someone **call collect** (609-951-1756) or write to the DANTES Program Office (see address below).

If you find that your college *accepts* DANTES credits, but does not *administer* the tests, here's what you do —

- Write to the DANTES Program Office for a list of colleges that administer the test. Chances are there's one nearby.
- Call any of the ETS Regional Offices (Tucker, GA; Oakland, CA; Evanston, IL; Phoenix, AZ; and Princeton, NJ. DANTES tests are given at all of those locations.
- In the military or in the reserves? DoD employees and family members may be able to take DANTES tests at a Military Education Office or Navy Campus Office.

Check Out the DANTES Test List.

→

Match the titles with the things you have learned. Consider how much college credit you might gain by taking some DANTES tests.

Your college should have fact sheets with more information about each DANTES test. If not, please write to us directly, or e-mail us at dantes@chauncey.com. Tell us which subjects interest you.

DANTES Program Office
Educational Testing Service
Princeton, NJ 08541-0001

MATHEMATICS

Introductory College Algebra
- Functions and graphs
- Fundamental algebraic operations
- Linear equations, inequalities, word problems
- Radicals and exponents

Principles of Statistics
- Descriptive statistics
- Correlation, regression, probability
- Chance models, sampling, test of significance

SOCIAL SCIENCE

Art of the Western World
- Greece and Rome
- Romanesque and Gothic
- Renaissance
- Baroque, rococo, neoclassicism, romanticism
- Twentieth century

The Civil War and Reconstruction
- Causes of the war
- Political situations and the war between 1861 and 1865
- Reconstruction

Contemporary Western Europe
- Historical background
- Political systems
- Societal issues
- Integration

Criminal Justice
- Criminal behavior
- Police and criminal justice system
- Court system
- Corrections

Drug and Alcohol Abuse
- Pharmacological principles
- Classification of drugs
- Properties of alcohol, sedative hypnotics, narcotic analgesics, stimulants, hallucinogens and other drugs
- Prevention/treatment

Foundations of Education
- Contemporary issues in education
- Past and current influences on education
- Interrelationships between contemporary issues and influences, past or current, in education

Fundamentals of Counseling
- Counselor roles and functions
- Counseling relationships
- Theoretical approaches
- Social and cultural foundations

General Anthropology
- Anthropology branches, methodologies, and theoretical perspectives
- Physical anthropology and archeology
- Nature of culture
- Social, economic, political, and religious organization

A History of the Vietnam War
- The first Vietnam War (1945-1954)
- LBJ goes to war (1964-1965)
- Tet Offensive (1968)
- Vietnamizing the war (1968-1973)
- Peace is at hand (1968-1973)

Human/Cultural Geography
- Basic facts and concepts
- Culture and environment
- Spatial processes
- Regional geography

Introduction to Law Enforcement
- U.S. criminal justice and police systems
- Issues and organization of police
- Constitutional law and precedents

Introduction to the Modern Middle East
- Geography and early history
- Nineteenth and early twentieth century
- First and Second World wars
- Soviet-U.S. rivalry, Suez Crisis, Iranian Revolution, Persian Gulf Crisis

Lifespan Developmental Psychology
- Biological development
- Perception, learning, and memory
- Social, emotional, and personality development
- Cognition and language

BUSINESS

Business Law II
- Sales of goods
- Debtor and creditor relations
- Business organizations
- Property
- Commercial paper

Business Mathematics
- Basic computation — fractions, percents, ratios, graphs
- Business applications

Introduction to Business
- Economic, international, government, and ownership issues
- Management process
- Human resources and production
- Marketing, finance

Introduction to Computers with Programming in BASIC
- Hardware, software, data management
- Communications and connectivity
- BASIC: I/O, expressions, loops, conditionals, strings, arrays

Management Information Systems
- Systems theory, analysis and design of systems
- Hardware and software
- Database management
- Telecommunications
- Management of the MIS functional area, informational support

Money and Banking
- Role of commercial banks and other intermediaries
- Federal Reserve System
- Macroeconomic activity, U.S. policy, international system

Personal Finance
- Financial goals, budgeting, cash management, economic terminology
- Credit, debt, and major purchases
- Taxes and insurance
- Investments, retirement and estate planning

Personnel/Human Resource Management
- Employment actions
- Training and development
- Performance appraisals
- Compensation and personnel legislation
- Labor relations

Principles of Finance
- Financial statements and ratio analysis
- Capital budgeting
- Time value of money
- Break-even and leverage, common stock

Principles of Financial Accounting
- General concepts and principles
- Financial statements
- Accruals and deferrals
- Current accounts

Principles of Supervision
- Roles and responsibilities
- Management functions
- Legal issues, stress management, unions, productivity and quality concerns

Organizational Behavior
- Individual and group processes and characteristics
- Organizational processes and behavior
- Change and development processes

APPLIED TECHNOLOGY

Technical Writing
- Elements of technical reports
- Style, grammar, mechanics, organizing
- Types of technical reports

HUMANITIES

Ethics in America
- Ethical traditions
- Analysis of issues in relationships
- Tradition versus analysis of issues

Introduction to World Religions
- Dimensions and approaches
- Primal religions, Hinduism, Buddhism, Confucianism, Taoism, Judaism, Christianity, Islam

Principles of Public Speaking
- Topics and purposes
- Organization and supporting materials
- Language, style, and delivery
- Audience analysis and adaptation

PHYSICAL SCIENCE

Astronomy
- The solar system
- The sun and stars
- Our galaxy and the universe
- Measurement

Environment and Humanity: The Race to Save the Planet
- Ecological concepts
- Human impact on the environment
- Environmental management
- Political processes and the future

Here's to Your Health
- Mental health and behavior
- Human development
- Substance use and abuse
- Fitness and nutrition
- Risk factors and disease
- Consumer awareness, environmental concerns

Physical Geology
- Earth materials
- Surface processes
- Internal earth processes
- Applications

Principles of Physical Science I
- Physics: Newton's laws, energy, thermodynamics optics, electricity, and magnetism
- Chemistry: matter, atomic theory and structure, chemical reactions

NOTE:

MOST SCHOOLS DO NOT ACCEPT ALL TESTS FOR ALL PROGRAMS. CHECK WITH THE SCHOOL FROM WHICH YOU EXPECT TO RECEIVE CREDIT.

DANTES

CONTENTS

from the official announcement for instructional purposes

DANTES Subject Standardized Tests (DSSTs)
Receive College Credit for Today's Educational Alternatives

Today, there are many educational alternatives to the classroom—you can learn from your job, your reading, your independent study, and special interests you pursue. You may already have learned the materials covered by some college-level courses.

The DANTES Subject Standardized Tests (DSSTs) is a nationally recognized testing program that gives you the opportunity to receive college credit for learning acquired outside the traditional college classroom. The program, developed by The Chauncey Group International, a subsidiary of Educational Testing Service (ETS), is administered year-round by colleges and universities throughout the United States. Thousands of individuals, particularly adults who are interested in boosting their careers, take the DSSTs annually and receive college credit. MAKE IT A POINT to take advantage of the DSST testing program; it speeds the educational process and provides the flexibility adults need, making earning a degree more feasible.

No prerequisites are required to take the DSSTs; however, because requirements differ from college to college, please check with the school where you are enrolled before taking any DSST. Over 1,200 colleges and universities currently award credit for DSSTs, and the number is growing every day. You can choose from over 35 test titles in the areas of Social Science, Business, Mathematics, Applied Technology, Humanities, and Physical Science. A brief description of each examination is found on pages 6 through 8 of this booklet.

Reach Your Career Goals Through Education

Use DSSTs to help you earn your degree, get a promotion, or simply demonstrate that you have college-level knowledge in subjects relevant to your work.

Save Time...

You don't have to sit through classes when you already know through your experiences most of what is being taught and can learn the rest yourself. You might be able to bypass introductory-level courses in subject areas you already know.

Save Money...

DSSTs save you money because the classes you bypass by earning credit through the credit-by-examination program are classes you won't have to pay for on your way to earning your degree. You can use the money instead to take more advanced courses that can be more challenging and rewarding.

Improve Your Chances for College Admission

Each college has its own admission policies; however, having passing scores for DSSTs on your transcript can provide strong evidence of how well you can perform at the college level.

Gain Confidence Performing at a College Level

Many adults returning to college find that lack of confidence is often the greatest hurdle to overcome. DSSTs can show you how successfully you can compete in the college environment because your performance is compared with the performance of college students.

Make Up for Courses You Might Have Missed

You may be ready to graduate from college and find that you are a few credits short of earning your degree. By using semester breaks, vacation time, or leisure time to study independently, you can prepare to take one or more DSSTs and graduate on time.

If You Cannot Attend Regularly Scheduled Classes...

If your lifestyle or responsibilities prevent you from attending regularly scheduled classes, you can earn your college degree from a college offering an external degree program. This type of program allows you to earn your degree by study and experience outside the traditional classroom. Credit is earned by passing credit-by-examinations such as the DSSTs, portfolio assessment, on-line classes, and other distance learning. To learn more, write or call:

Charter Oak College
66 Cedar Street
Newington, Connecticut 01611
(203) 666-4595

Regents College
7 Columbia Circle
Albany, New York 12203-5159
(518) 464-8500

Thomas Edison State College
101 West State Street
Trenton, New Jersey 08608
(609) 984-1150

Other colleges and universities offer external degree or distance learning programs. For additional information, contact the college you plan to attend or:

Center for Adult Learning and Education Credentials
American Council on Education
1 DuPont Circle
Washington, DC 20036
(202) 939-9475

Where to Take DSSTs

DANTES Subject Standardized Tests (DSSTs) are administered at more than 600 locations nationwide. Each location determines the frequency and scheduling of test administrations.

Your college or one near you may be a DSSTs test site. The free booklet DSST Participating Colleges and Universities lists all DSST test sites and their addresses and phone numbers so you can contact them about scheduling testing. In addition to listing DSST test sites, DSST Participating Colleges and Universities also lists colleges that award DSST credits.

DSST Participating Colleges and Universities is available by writing: DANTES Program, The Chauncey Group International, 664 Rosedale Road, Princeton, New Jersey 08540. You may also order by e-mail at DANTES@Chauncey.com. Please be sure to include your complete name and mailing address.

Earning Credit for DSST Examinations

Credit is awarded for passing scores on DSST exams at more than 1,200 colleges and universities in the United States.

To find out whether or not the college of your choice awards credit for passing DSST scores, contact the admissions office or counseling and testing office. The college can also provide information on the scores required for awarding credit, the number of credit hours awarded, and any courses that can be bypassed with satisfactory scores.

It is important that you contact the college of your choice as early as possible because colleges' credit-awarding policies differ.

Arranging for DSST test administration

After you have selected a college that administers DSSTs, you will need to contact their representative listed in the DSST Participating Colleges and Universities to make arrangements for date and time of testing. Please be aware that certain colleges provide DSST testing services to enrolled students only.

The fee to take a DSST is $27 per test. This fee entitles you to two score reports after the test is scored. One will be sent directly to you and the other will be sent directly to the college or university of your choice. You may pay the fee with a certified check or money order made payable to DANTES Program or you may charge the fee to your Visa, MasterCard, or American Express credit card.

In addition, the test site may also require a test administration fee for each examination, to be paid directly to the institution. Contact the test site to determine its fee and payment policy.

Other Testing Arrangements

If you are unable to find a test site near your location, you may want to contact the testing office of a local accredited college or university to determine whether a representative from that office will agree to administer the test(s) to you on a one-time only basis. The school's representative should then contact the DANTES Client Services Representative at (609) 720-6532 to arrange for this special permission.

If you are a civilian at or near a military base and you do not have access to a civilian test center, you may want to contact the base education officer to see whether security and space permit you to test on the base. This decision is based on the discretion of the base representative.

Testing Accommodations for Students with Disabilities

The Chauncey Group International (Chauncey Group) is committed to serving test takers with disabilities by providing services and reasonable testing accommodations as set forth in the provisions of the Americans with Disabilities Act (ADA). If you have a disability, as prescribed by the ADA, and require special testing services or arrangements, please contact the test administrator at the test site. You will be asked to submit to the test administrator documentation of your disability and your request for special accommodations. The test administrator will then forward your documentation along with your request for testing accommodations to the Chauncey Group for approval. Please submit your request as well in advance of your test date as possible so that we may have the opportunity to make the necessary accommodations.

Only test takers with documented disabilities are eligible for special accommodations.

Description of the DSST Examinations

Mathematics

- **Fundamentals of College Algebra** covers mathematical concepts such as fundamental algebraic operations; linear, absolute value; quadratic equations, inequalities, radials, exponents and logarithms, factoring polynomials and graphing.
- **Principles of Statistics** tests the understanding of the various topics of statistics, both qualitatively and quantitatively, and the ability to apply statistical methods to solve a variety of problems. The topics included in this test are descriptive statistics; correlation and regression; and probability.

Social Science

- **Art of the Western World** deals with the history of art during the following periods: classical; Romanesque and Gothic; early Renaissance; high Renaissance, Baroque; rococo; neoclassicism and romanticism; realism, impressionism and post-impressionism; early twentieth century; and post-World War II.
- **Contemporary Western Europe: 1946 - 1990** tests the knowledge of basic facts and terms and the understanding of concepts and principles related to the areas of the historical background of the aftermath of the Second World War and rebuilding of Europe, national political systems, and issues in Western European societies.
- **An Introduction to the Modern Middle East** emphasizes core knowledge (including geography, Judaism, Christianity, Islam, ethnicity); nineteenth-century European impact; twentieth-century Western influences; World Wars I and II; new nations and Soviet-United States rivalry; the Suez Crisis; the Iranian Revolution; the Soviet invasion of Afghanistan, and the Persian Gulf Crisis.
- **Human/Cultural Geography** includes the Earth and basic facts (coordinate systems, maps, physiography, atmosphere, soils and vegetation, water); culture and environment, spatial processes (social processes, modern economic systems, settlement patterns, political geography); and regional geography.
- **Rise and Fall of the Soviet Union** covers Russia under the Old Regime; the Revolutionary Period; New Economic Policy; Pre-war Stalinism; The Second World War; Post-war Stalinism; The Krushchev Years; The Brezhnev Era; and reform and collapse.
- **A History of the Vietnam War** covers the history of the roots of the Vietnam War; the First Vietnam War (1946-1954); pre-war developments (1954-1963); American involvement in the Vietnam War; Tet (1968); Vietnamizing the War (1968-1973); Cambodia and Laos; peace; legacies and lessons.
- **The Civil War and Reconstruction** covers the Civil War from pre-secession (1861) through Reconstruction. It includes causes of the war; secession; Fort Sumter; the war in the east and in the west; major battles; the political situation; assassination of Lincoln; end of the Confederacy; and Reconstruction.
- **Foundations of Education** includes topics such as contemporary issues in education; past and current influences on education (philosophies, democratic ideals, social/economic influences); and the interrelationships between contemporary issues and influences.
- **Life-span Developmental Psychology** covers models and theories; methods of study; ethical issues; biological development; perception, learning and memory; cognition and language; social, emotional, and personality development; social behaviors, family life cycle, extrafamilial settings; singlehood and cohabitation; occupational development and retirement; adjustment to life stresses; and bereavement and loss.
- **Drug and Alcohol Abuse** includes such topics as drug use in society; classification of drugs; pharmacological principles; alcohol (types, effects of, alcoholism); general principles and use of sedative hypnotics, narcotic analgesics, stimulants, and hallucinogens; other drugs (inhalants, steroids); and prevention/treatment.
- **General Anthropology** deals with anthropology as a discipline; theoretical perspectives; physical anthropology; archaeology; social organization; economic organization; political organization; religion; and modernization and application of anthropology.
- **Introduction to Law Enforcement** includes topics such as history and professional movement of law enforcement; overview of the U.S. criminal justice system; police systems in the U.S.; police organization, management, and issues; and U.S. law and precedents.

- **Criminal Justice** deals with criminal behavior (crime in the U.S., theories of crime, types of crime); the criminal justice system (historical origins, legal foundations, due process); police; the court system (history and organization, adult court system, juvenile court, pre-trial and post-trial processes); and corrections.

- **Fundamentals of Counseling** covers historical development (significant influences and people); counselor roles and functions; the counseling relationship; and theoretical approaches to counseling.

Business

- **Principles of Finance** deals with financial statements and planning; time value of money; working capital management; valuation and characteristics; capital budgeting; cost of capital; risk and return; and international financial management.

- **Principles of Financial Accounting** includes topics such as general concepts and principles, accounting cycle and classification; transaction analysis; accruals and deferrals; cash and internal control; current accounts; long- and short-term liabilities; capital stock; and financial statements.

- **Personnel/Human Resource Management** covers general employment issues; job analysis; training and development; performance appraisals; compensation issues; security issues; personnel legislation and regulation; labor relations and current issues.

- **Organizational Behavior** deals with the study of organizational behavior (scientific approaches, research designs, data collection methods); individual processes and characteristics; interpersonal and group processes and characteristics; organizational processes and characteristics; and change and development processes.

- **Principles of Supervision** deals with the roles and responsibilities of the supervisor; management functions (planning, organization and staffing, directing at the supervisory level); and other topics (legal issues, stress management, union environments, quality concerns).

- **Business Law II** covers topics such as sales of goods; debtor and creditor relations; business organizations; property; and commercial paper.

- **Introduction to Computers with Programming** in Basic includes topics such as history and technological generations; hardware/software; applications to information technology; program development; data management; communications and connectivity; computing and society; and programming in BASIC.

- **Management Information Systems** covers systems theory, analysis and design of systems, hardware and software; database management; telecommunications; management of the MIS functional area and informational support.

- **Introduction to Business** deals with economic issues affecting business; international business; government and business; forms of business ownership; small business, entrepreneurship and franchise; management process; human resource management; production and operations; marketing management; financial management; risk management and insurance; and management and information systems.

- **Money and Banking** covers the role and kinds of money; commercial banks and other financial intermediaries; central banking and the Federal Reserve system; money and macroeconomics activity; monetary policy in the U.S.; and the international monetary system.

- **Personal Finance** includes topics such as financial goals and values; budgeting; credit and debt; major purchases; taxes; insurance; investments; and retirement and estate planning.

- **Business Mathematics** deals with basic operations with integers, fractions, and decimals; round numbers; ratios; averages; business graphs; simple interest; compound interest and annuities; net pay and deductions; discounts and markups; depreciation and net worth; corporate securities; distribution of ownership; and stock and asset turnover.

Physical Science

- **Astronomy** covers the history of astronomy, celestial mechanics; celestial systems; astronomical instruments; the solar system; nature and evolution; the galaxy; the universe; determining astronomical distances; and life in the universe.

- **Here's to Your Health** covers mental health and behavior; human development and relationships; substance abuse; fitness and nutrition; risk factors, disease, and disease prevention; and safety, consumer awareness, and environmental concerns.

- **Environment and Humanity** deals with topics such as ecological concepts (ecosystems, global ecology, food chains and webs); environmental impacts; environmental management and conservation; and political processes and the future.

- **Principles of Physical Science I** includes physics: Newton's Laws of Motion; energy and momentum; thermodynamics; wave and optics; electricity and magnetism; chemistry: properties of matter; atomic theory and structure; and chemical reactions.

- **Physical Geology** covers Earth materials; igneous, sedimentary, and metamorphic rocks; surface processes (weathering, groundwater, glaciers, oceanic systems, deserts and winds, hydrologic cycle); internal Earth processes; and applications (mineral and energy resources, environmental geology).

Applied Technology

- **Technical Writing** covers topics such as theory and practice of technical writing; purpose, content, and organizational patterns of common types of technical documents; elements of various technical reports; and technical editing. Students may write a short essay on one of the technical topics provided. The essay will not be scored by the Chauncey Group; however, a copy of the essay will be provided with the score report or transcript.

Humanities

- **Ethics in America** deals with ethical traditions (Greek views, Biblical traditions, moral law, consequential ethics, feminist ethics); ethical analysis of issues arising in interpersonal and personal-societal relationships and in professional and occupational roles; and relationships between ethical traditions and the ethical analysis of situations. Students may write an essay to analyze a morally problematic situation in terms of issues relevant to a decision and arguments for alternative positions. The essay will not be scored by the The Chauncey Group; however, a copy of the essay will be provided with the score report or transcript.

- **Introduction to World Religions** covers topics such as dimensions and approaches to religion; primal religions; Hinduism; Buddhism; Confucianism; Taoism; Judaism; Christianity; and Islam.

- **Principles of Public Speaking** consists of two parts: Part One requires the student to record an impromptu persuasive speech that will be scored. Part Two consists of multiple-choice questions covering considerations of public speaking; audience analysis; purposes of speeches; structure/organization; content/supporting materials; research; language and style; delivery; communication apprehension; listening and feedback; and criticism and evaluation.

HOW TO ORDER FACT SHEETS/STUDY GUIDES

For each test, there is a Fact Sheet/Study Guide that gives information about the topics that will be covered, books that would be useful for review, the number of questions on the test, and the passing score recommended by the American Council on Education (ACE) for granting credit. It is recommended that students check with the school where they are enrolled about the school's requirements for earning credit. Some schools set higher standards than the ACE minimum recommendation. The Fact Sheet/Study Guides also contain sample test questions to give you an idea of what the test is like. You may order a free copy of the Fact Sheet/Study Guide you need by completing and mailing in the Fact Sheet/Study Guide Order Form on page 11 or sending an e-mail request to DANTES@Chauncey.com. When sending an e-mail request, please include your name and complete mailing address.

ON THE DAY OF THE EXAMINATION

Here are a few tips to help you on the day of your examination:

- Arrive on time as a courtesy to the test administrator. Remember, many test administrators have responsibilities outside the testing office and may have a specific amount of time allotted for your test administration.

- Have your personal photo identification with you. <u>Anyone who fails to present a photo I.D. will not be tested.</u>

- Bring several No.2 (soft-lead) pencils with good erasers, a watch, and a black pen if you will be writing an essay.

- Do not bring books or papers.

- Do not bring an alarm watch that beeps, a telephone, or a phone beeper into the testing room.

- The use of non-programmable calculators, slide rules, scratch paper and/or other materials is permitted for some of the tests. Please refer to the section entitled Auxiliary Test Materials on page 15 for specific information.

DANTES SUBJECT STANDARDIZED TESTS SCORING POLICIES

Your DSST examination scores are reported only to you, unless you request that they be sent elsewhere. If you want your scores sent to a college you must grid the correct code number or write the name and address of the school on your answer sheet at the time you take the test.

Receiving Your Score Report

DSST answer sheets are scanned by computer. Allow four weeks after testing to receive your score report.

Do not call before the allotted time for scoring has elapsed—this will not expedite the processing of your scores. Please note that scores will not be reported to students over the telephone because of privacy and security considerations.

Scoring of Public Speaking Speech and Retention of Essays

The speech portion of Public Speaking will be sent to speech raters who are faculty members at accredited colleges that currently teach the course. These raters have received special training to ensure uniformity. Because of this, scores for the Public Speaking examination are normally available in 6 - 8 weeks. If you take the Public Speaking examination and fail, you may retest after 6 months (180 days) on a different speech topic. You must wait 2 years to re-test using the same speech topic.

The essays for Ethics in America and Technical Writing are optional and thus are not scored by raters. The essays are forwarded to the colleges and universities that you designate, along with your score report, for their use in determining the award of credit.

How to Get Transcripts

To request a transcript, use the form located on page 13. There is a $10 fee for each transcript you request.

Availability/Use of Scores

College score requirements for awarding credit vary. Please contact the college or university in which you are enrolled for information on the minimum score it requires to grant credit.

Candidate scores that have been released to colleges may be used for educational research or validity studies by the receiving colleges or by The Chauncey Group International, but the names of individuals will not be revealed or attached to particular scores.

Score Validity

The DANTES Program is obligated to report scores that accurately reflect the performance of the test taker. The program maintains test administration and test security standards designed to assure that all test takers are given the same opportunity to demonstrate their abilities and to prevent any test taker from gaining an unfair advantage.

The DANTES Program reserves the right to cancel any test score if the test taker engages in misconduct, if there is a testing irregularity, or if there is reason to question the score's validity.

Reviews of questionable scores are confidential. If it is necessary to cancel scores that have already been reported, score users are notified, but the reasons for cancellation are not disclosed.

FREQUENTLY ASKED QUESTIONS

How should I study for the examinations?

For each of the DSSTs there is a Fact Sheet/Study Guide which outlines the topics covered by each test. Each Fact Sheet/Study Guide includes a list of sample questions and a list of recommended references for you to use. You may order a free copy of the Fact Sheet/ Study Guides you need by completing and mailing the Fact Sheet/Study Guide Order Form on page 11 or by sending an e-mail request to DANTES@chauncey.com. Please include your name and complete mailing address.

In order to pass the test, must I study from one of the recommended references?

The recommended references are a listing of books that were currently being used as text books for the same course title at the time the test was written. If some time has passed since then, you may have difficulty finding the exact title or edition of the book. You might want to consider visiting the bookstore of a local college or university and comparing the text book they currently use for the same class title to the topics listed on the Fact Sheet/Study Guide. If the topics listed on the Fact Sheet/Study Guide are covered by the selected textbook, then you should be reasonably prepared for the test by studying that text book.

Is there a penalty for guessing on the tests?

There is no penalty for guessing on the DANTES Subject Standardized Tests (DSSTs), so you should mark an answer for each question.

How much time will I have to complete the test?

Many DSSTs can be completed within 90 minutes; however, additional time can be allowed if necessary.

What should I do if I find an error in a test question?

If you think there is an error in a test question or any aspect of a test administration, continue with the test, but report the information to the testing supervisor. This may be done after the test by asking that it be noted on the Supervisor's Irregularity Report or by writing to DANTES Program, The Chauncey Group International, Princeton, NJ 08540 and indicating the form and question number(s) or circumstances as well as your name and address.

Will my test scores be released without my permission?

Your test score will not be released to anyone other than the school you designate on your answer sheet unless you write to us and ask us to send a transcript elsewhere. Instructions about how to do this can be found on your score report. Your scores may be used for research purposes, but individual scores are never made public nor are individuals identified if research findings for a group are made public.

If I do not achieve a passing score on the test, how long must I wait until I can take the test again?

If you do not receive a score on the test that will enable you to obtain credit for the course, you may take the test again after six months (180 days). Please do not attempt to take the test before six months (180 days) have passed because you will receive a score report marked invalid and will have wasted the time and the money you invested in retesting. There must be 180 days between test dates in order for your score report to be valid.

Can my test scores be canceled?

The test administrator is required to report any irregularities to the DANTES Program. The consequence of bringing unauthorized materials into the testing room or giving or receiving help will be the forfeiture of your test fee. The DANTES Program reserves the right to cancel scores and not issue score reports in such situations.

What can I do if I feel that my test scores were not accurately reported?

If you have reason to believe that your score(s) were not accurately reported, you may request to have your answer sheet reviewed and hand scored. There is a $10 fee for this service. Please submit a check or money order for $10 payable to DANTES Program with your request in writing for score verification to: DANTES Program, P.O. Box 6604, Princeton, NJ 08541-6604. Include your full name, the test title, the date you took the test, and your Social Security number. We will notify you of the results of hand scoring and, if your scores were not accurate, a corrected score report will be forwarded to you and to any school to which score reports were sent. In the event that a scoring error occurred, your $10 fee will be refunded.

What does ACE recommendation mean?

The ACE recommendation is the minimum passing score recommended by the American Council on Education for any given test. It is equivalent to the average score of students in the DANTES norming sample who received a grade of C for the course. Some schools require a score higher than the ACE recommendation.

How long are my test scores kept on file?

Test scores are kept on file for a period of 20 years.

REMEMBER TO MAKE IT A POINT

You are always learning, whether it be inside a classroom, on your job, or running a household. You can gain experience through retail sales experience, parenting, reading, hobbies (such as Toastmaster's), and non-credit business training. MAKE IT A POINT to get credit for what you already know with DANTES Subject Standardized Tests.

DANTES SUBJECT STANDARDIZED TESTS (DSSTs)

FACT SHEET ❖ ORDER FORM

INSTRUCTIONS:

Please ship the quantities of the DANTES Fact Sheets on the reverse side to:

Name ______________________________

Address ______________________________

City ______________________ **State** _____ **Zip Code** __________

Today's Date ____________________

To place your order, please mail the completed form to:

The DANTES Program
The Chauncey Group International
PO Box 6604, MS-23P
Princeton, NJ 08541-6604

Fax: (609) 720-6800
For confirmation of receipt of FAX, call: 609-720-6726

For information on The DANTES Program, call (609) 720-6740.

Fact Sheets for DANTES Subject Standardized Tests

Quantity	*Title*
	MATHEMATICS
______	Fundamentals of College Algebra
______	Principles of Statistics
	SOCIAL SCIENCE
______	Art of the Western World
______	Contemporary Western Europe: 1946-1990
______	An Introduction to the Modern Middle East
______	Human/Cultural Geography
______	Rise and Fall of the Soviet Union
______	A History of the Vietnam War
______	The Civil War and Reconstruction
______	Foundations of Education
______	Lifespan Developmental Psychology
______	General Anthropology
______	Drug and Alcohol Abuse
______	Introduction to Law Enforcement
______	Criminal Justice
______	Fundamentals of Counseling
	BUSINESS
______	Principles of Finance
______	Principles of Financial Accounting
______	Personnel/Human Resource Management
______	Organizational Behavior
______	Principles of Supervision
______	Business Law II
______	Introduction to Computers with Programming in BASIC
______	Introduction to Business
______	Money and Banking
______	Personal Finance
______	Management Information Systems
______	Business Mathematics

Quantity	*Title*
	PHYSICAL SCIENCE
______	Astronomy
______	Here's to Your Health
______	Environment and Humanity: The Race to Save the Planet
______	Principles of Physical Science I
______	Physical Geology
	APPLIED TECHNOLOGY
______	Technical Writing
	HUMANITIES
______	Ethics in America
______	Introduction to World Religions
______	Principles of Public Speaking

DANTES SUBJECT STANDARDIZED TESTS **(DSSTs)**

CIVILIAN TRANSCRIPT REQUEST FORM FOR DSST SCORES

**TRANSCRIPT REQUEST FORM
FOR DSST SCORES**

To avoid a delay, do not send a letter with this form.

FEE:

A **$10** fee is charged for **each** transcript requested. A transcript may include any or all DSST scores.

Enclose a check or money order payable to DANTES Program. If you prefer to pay for your transcript with a credit card, please supply the information below. Your account will be charged with the total amount due. Incomplete forms or forms received without fees will be returned to the requester. **DO NOT SEND CASH OR STAMPS.**

SEND COMPLETED FORM TO:

DANTES Program
The Chauncey Group International
P.O. Box 6604
Princeton, NJ 08541-6604

☐ Check or Money Order enclosed (In U.S. Dollars) Please charge my ☐ MasterCard ☐ Visa ☐ American Express

My credit card number is: Expiration date: ____ / ____ (Month / Year)

__ __ __ __ - __ __ __ __ - __ __ __ __ - __ __ __ __

PLEASE PRINT ALL INFORMATION REQUESTED BELOW.

Last Name (Include Maiden Name or Former Last Names) First MI / Social Security Number

Present Address Number and Street / Date of Birth

City and State Zip Code Phone Number

I understand that my signature permits DANTES to release my transcript to the designated recipient(s).

(Signature)

Please prepare my transcript and include the following (check only one):

☐ Scores on all tests ☐ Only test scores that are at or above the ACE Recommended Minimum Score ☐ Only scores on tests named below

TEST TITLES:

Approximate Date of Last DANTES Test

☐ Please send a copy of my transcript to the following school(s) (**$10.00 per transcript**):

School Name	School Name
Attn:	Attn:
Address	Address
City & State Zip Code	City & State Zip Code

Notes for Test Takers

Here is some information about the DANTES test you are planning to take that you may find helpful.

Preparing for the Test

For each test, there is a fact sheet that gives information about the topics that will be covered, books that would be useful for review, the number of questions on the test, and the minimum score recommended by the American Council on Education (ACE) for credit. (You should check the passing score with your school; some have set higher standards than the ACE minimum recommendation.) Some of the fact sheets also contain sample test questions to give you an idea of what the test is like. Copies of these sheets are available without charge from the test administrator, your counselor, by phone at 609-720-6740 or by e-mail at DANTES@chauncey.com.

Guessing

There is no penalty for guessing on the DANTES Subject Standardized Tests, so you should mark an answer for each question.

Test Timing

These tests are untimed. It generally takes about 1 1/2 hours to take a DANTES Subject Standardized Test; however, you may work longer if necessary.

Test Materials

The use of non-programmable calculators, slide rules, scratch paper and/or other materials is permitted for some of the tests. The test administrator has information about whether the test(s) you are planning to take fall into this category and which materials are permitted. If you wish to bring any materials into the test room, please check before testing to be sure they are permitted.

DANTES

Subject Standardized Tests

Test Administration

If you think there is an error in a test question, continue with the test, but report the information to The Chauncey Group International. This may be done by telling the supervisor after the test, asking that it be noted on the Supervisor's Irregularity Report, by writing to **The Chauncey Group, DANTES Program, CN-6604, Princeton, NJ 08541-6604**, indicating the form and question numbers as well as your name and address.

Confidentiality

Your test score on your test will not be released to anyone other than the school you designate on your answer sheet unless you write to us and ask us to send a transcript elsewhere. Instructions about how to do this will be found on your score report. Your scores may be used for research purposes, but individual scores are never made public nor are individuals identified if research findings for a group are made public.

Students with Disabilities

Special testing arrangements are authorized for students with disabilities. If you are in need of this service, speak to your test administrator in advance of the testing. A reader or writer may be used or other special arrangements may be made to minimize the effect of your disability on your score.

Retesting

If you do not receive a score on the test that will enable you to obtain credit for the course, you may take the test again after six months (180 days). Please do not attempt to retake the test before six months (180 days) have elapsed because you will receive a score report marked invalid and will have wasted the time and money you invested in retesting.

Score Cancellation

The test administrator is required to report any irregularities to The Chauncey Group. The consequences of bringing unauthorized materials into the testing room or giving or receiving help will be the forfeiture of your test fee. The Chauncey Group reserves the right to cancel scores and not issue score reports in such situations.

Score Verification

If you have reason to believe that your score(s) were not accurately reported, you may request to have your answer sheet hand scored. There is a $10 fee for this service. Please submit a check or money order for $10 payable to **DANTES Program** with your request in writing for score verification to: **The Chauncey Group International, P.O. Box 6604, Princeton, NJ 08541-6604**. Include your full name, the test title, the date you took the test and your Social Security number. We will notify you of the results of hand scoring and, if your scores were not accurate, a corrected score report will be forwarded to you and any school, to which score reports were sent. In the event that a Chauncey Group error occurred, your money will be refunded.

DANTES

Subject Standardized Tests

ACE Credit Recommendations

American Council on Education
Credit Recommendations

Test Form Number	Title	Credit Amount in Semester Hours	Minimum Score
MATHEMATICS			
SF/SG 424	Fundamentals of College Algebra	3B	47
SE 450	Principles of Statistics	3B	48
SOCIAL SCIENCE			
SE 461	Art of the Western World	3B	48
SE 465	Contemporary Western Europe: 1946-1990	3B	48
SE 469	An Introduction to the Modern Middle East	3B	44
SF/SG 470	Human/Cultural Geography	3B	48
SE 471	Rise and Fall of the Soviet Union	3BU	45
SF 473	A History of the Vietnam War	3B	49
SE/SF 483	The Civil War and Reconstruction	3BU	47
SE 489	Foundations of Education	3B	46
SF/SG 490	Lifespan Developmental Psychology	3B	46
SF 494	General Anthropology	3B	47
SE 495	Drug and Alcohol Abuse	3BU	49
SG/SH 497	Introduction to Law Enforcement	3B	45
SF/SG 498	Criminal Justice	3B	49
SG/SH 562	Fundamentals of Counseling	3B	45
BUSINESS			
SF/SG 524	Principles of Finance	3BU	46
SF 525	Principles of Financial Accounting	3B	49
SF 530	Personnel/Human Resource Management	3B	48
SF/SG 531	Organizational Behavior	3B	48
SE/SF 532	Principles of Supervision	3B	46
SE 534	Business Law II	3BU	52
SG/SH 536	Introduction to Computing	3B	45
SE/SF 543	Introduction to Business	3B	46
SG/SH 548	Money and Banking	3BU	48
SE 550	Personal Finance	3B	46
SE 551	Management Information Systems	3BU	46
SF/SG 812	Business Mathematics	3B	48

Test Form Number	Title	Credit Amount in Semester Hours	Minimum Score
PHYSICAL SCIENCE			
SF 500	Astronomy	3B	48
SF/SG 508	Here's to Your Health	3B	48
SF/SG 511	Environment and Humanity: The Race to Save the Planet	3B	46
SE 512	Principles of Physical Science I	3B	47
SF 519	Physical Geology	3B	46
APPLIED TECHNOLOGY			
SF/SG 820	Technical Writing *(TWE Technical Writing Essay)*	3B	46
HUMANITIES			
SF/SG 474	Ethics in America *(SF 474 Ethics in America Essay)*	3B	46
SE 496	Introduction to World Religions	3B	49
SE/SF 815*	Principles of Public Speaking	3B	47

*In addition to a minimum score of 47 on the multiple-choice test, an examinee must also receive a passing grade on the speech.

B = Baccalaureate Program
BU = Baccalaureate Upper Division

Fact Sheet

DANTES Subject Standardized Tests

HUMAN/CULTURAL GEOGRAPHY

TEST INFORMATION

This test was developed to enable schools to award credit to students for knowledge equivalent to that which is learned by students taking the course. The school may choose to award college credit to the student based on the achievement of a passing score. The passing score for each examination is determined by the school based on recommendations from the American Council on Education (ACE). This minimum credit-awarding score is equal to the mean score of students in the norming sample who received a grade of C in the course. Some schools set their own standards for awarding credit and may require a higher score than the ACE recommendation. Students should obtain this information from the institution where they expect to receive credit.

CONTENT

The following topics, which are commonly taught in courses on this subject, are covered by this examination.

Approximate **Percent**

I. The Earth - basic facts and concepts — **23%**
- A. Coordinate systems, seasons, time
- B. Maps and cartography
- C. Physiography
  1. Tectonics
  2. Gradation, land forms
- D. Atmosphere
  1. Weather
  2. Climate
- E. Soils and vegetation
- F. Water

Approximate **Percent**

II. Culture and Environment — **40%**
- A. Cultural systems and processes
  1. Domestication of plants and animals
  2. Cultural origins and dispersals
  3. Language, religion, material culture
- B. Population
  1. Population characteristics and patterns
  2. Population growth and resource use
- C. Natural resources
  1. Human impact on the environment
  2. Environmental risks and hazards
  3. Development

III. Spatial Processes — **32%**
- A. Social processes
  1. Human spatial behavior
  2. Interaction through movement
  3. Diffusion
- B. Modern economic systems
  1. Agriculture and extractive activities
  2. Manufacturing
  3. Retailing, services, and high technology

from the official announcement for instructional purposes

A subsidiary of

te

Percent

C. Settlement patterns
 1. Principles of settlement and land use
 2. Rural settlement
 3. Urban settlement

D. Political Geography
 1. Territories and boundaries
 2. Politics and location

E. Social problems

IV. Regional Geography **5%**
 A. Defining a region
 B. Geopolitical regions

Questions on the test require candidates to demonstrate the following abilities. Some questions may require more than one of these abilities.

- Knowledge of basic facts and terms (about 55-65 percent of the examination)
- Understanding of concepts and principles (about 35-45 percent of the examination)
- Ability to apply knowledge to specific problems and situations (about 5-10 percent of the examination)

SAMPLE QUESTIONS

1. Which stage in the demographic transition is characterized by high, steady birth rates combined with decreasing death rates?

 (A) Rapid growth
 (B) Slow growth
 (C) No growth
 (D) Decline

2. A landscape that is pitted with depressions, has a stream passing underground into a cavern, and has several large underground caves is an example of which of the following?

 (A) Kettle and kame
 (B) Horst and graben
 (C) Karst
 (D) Outwash plain

3. Which of the following is equivalent to a scale of one inch equals one mile?

(A) 1: 13,360
(B) 1: 23,360
(C) 1: 63,360
(D) 1: 93,360

4. Which of the following is NOT an example of a centrifugal force in political geography?

(A) Basques in Spain
(B) Palestinians in Lebanon
(C) Quebecois in Canada
(D) African Americans in the United States

5. A well-known advocate of "hearths of domestication" for plants and animals is

(A) Richard Hartshorne
(B) Carl 0. Sauer
(C) Brian J.L. Berry
(D) Walter Christaller

6. Three types of world economic systems are

(A) socialism, communism, and democracy
(B) plantation, commercial, and planned
(C) subsistence, commercial, and planned
(D) shifting cultivation, swidden, and subsistence

7. According to the core-periphery model, a megalopolis is a

(A) core region
(B) upward-transition region
(C) resource-frontier region
(D) downward-transition region

8. All of the following are considered examples of west-coast desert EXCEPT the

(A) Namib Desert
(B) Gobi Desert
(C) Atacama Desert
(D) Great Sandy Desert

9. Which of the following is true of the population density in North American cities?
   (A) It is greatest at the center and gradually decreases toward the suburbs.
   (B) It gradually increases from the center to the suburbs.
   (C) It is low at the center, highest outside the center, with decreases toward the suburbs.
   (D) It is relatively uniform throughout the city.

10. Which of the following nation states is NOT a member of OPEC?
    (A) Nigeria
    (B) Saudi Arabia
    (C) Libya
    (D) Mexico

STUDYING FOR THE EXAMINATION

The following is a list of reference publications that were being used as textbooks in college courses of the same or similar title at the time the test was developed. Appropriate textbooks for study are not limited to those listed below. If you wish to obtain study resources to prepare for the examination, you may reference either the current edition of the following titles **or** textbooks currently used at a local college or university for the same class title. It is recommended that you reference **more than one textbook** on the topics outlined in this fact sheet. You should **begin by checking textbook content against the content outline** included on the front page of this Fact Sheet **before** selecting textbooks that cover the test content from which to study. Textbooks may be found at the campus bookstore of a local college or university offering a course on the subject.

Sources for study material suggested but not limited to the following:

Bergman, Edward, Tom L.McKnight. *Introduction to Geography.* Englewood Cliffs, NJ: Prentice Hall, current edition.

DeBlij, H.J., P.O. Muller. *Geography: Realms, Regions and Concepts.* New York, NY: Wiley and Sons, current edition.

DeBlij, H.J. P.O. Muller. *Human Geography: Culture, Society and Space.* New York, NY: Wiley and Sons, current edition.

Fellman, Jerome, A. Getis and J. Getis. *Human Geography: Landscape of Human Activities.* Boston, MA: WCB/McGraw Hill, current edition.

Gabler, Robert E., Robert J. Sager and Daniel L.Wise. *Essentials of Physical Geography.* Fort Worth, TX: Saunders College Publishing, current edition.

Goode, J. Paul. *Goode's World Atlas.* Chicago: Rand McNally, current edition.

Hartshorn, Truman A., John W. Alexander. *Economic Geography.* Englewood Cliffs, NJ: Prentice Hall, current edition.

McKnight, Tom L. *Physical Geography: A Landscape Appreciation.* Upper Saddle River, NJ: Prentice Hall, current edition.

Rubenstein, James H. *Cultural Landscape: Introduction to Human Geography.* Upper Saddle River, NJ: Prentice Hall, current edition.

Stansfield, Charles A. *Building Geographic Literacy.* Upper Saddle River, NJ: Prentice Hall, current edition.

Strahler, Alan H., Arthur N. Strahler. *Modern Physical Geography.* New York, NY: Wiley and Sons, current edition.

Current textbook used by a local college or university for a course on the subject.

CREDIT RECOMMENDATIONS

The Center For Adult Learning and Educational Credentials of the American Council on Education (ACE) has reviewed and evaluated the DSST examination development process. The American Council on Education has made the following recommendations:

Area or Course Equivalent:	Human/Cultural Geography
Level:	Lower-level baccalaureate
Amount of Credit:	Three (3) semester hours
Source:	ACE Commission on Educational Credit and Credentials

INFORMATION

Colleges and universities that would like additional information about the national norming, or assistance in local norming or score validation studies should write to: DSST Program, Mail Stop 11-P, The Chauncey Group International, 664 Rosedale Road, Princeton, New Jersey 08540.

It is advisable that schools develop a consistent policy about awarding credit based on scores from this test and that the policy be reviewed periodically. The Chauncey Group will be happy to help schools in this effort.

Correct Responses: 1.A; 2.C; 3.C; 4.D; 5.B; 6.C; 7.A; 8.B; 9.C; 10.D

I.N. D204296

HOW TO TAKE A TEST

You have studied hard, long, and conscientiously.

With your official admission card in hand, and your heart pounding, you have been admitted to the examination room.

You note that there are several hundred other applicants in the examination room waiting to take the same test.

They all appear to be equally well prepared.

You know that nothing but your best effort will suffice. The "moment of truth" is at hand: you now have to demonstrate objectively, in writing, your knowledge of content and your understanding of subject matter.

You are fighting the most important battle of your life—to pass and/or score high on an examination which will determine your career and provide the economic basis for your livelihood.

What extra, special things should you know and should you do in taking the examination?

BEFORE THE TEST

YOUR PHYSICAL CONDITION IS IMPORTANT

If you are not well, you can't do your best work on tests. If you are half asleep, you can't do your best either. Here are some tips:

1. Get about the same amount of sleep you usually get. Don't stay up all night before the test, either partying or worrying—DON'T DO IT.
2. If you wear glasses, be sure to wear them when you go to take the test. This goes for hearing aids, too.
3. If you have any physical problems that may keep you from doing your best, be sure to tell the person giving the test. If you are sick or in poor health, you really cannot do your best on any test. You can always come back and take the test some other time.

AT THE TEST

EXAMINATION TECHNIQUES

1. Read the *general* instructions carefully. These are usually printed on the first page of the examination booklet. As a rule, these instructions refer to the timing of the examination; the fact that you should not start work until the signal and must stop work at a signal, etc. If there are any *special* instructions, such as a choice of questions to be answered, make sure that you note this instruction carefully.

2. When you are ready to start work on the examination, that is as soon as the signal has been given, read the instructions to each question booklet, underline any key words or phrases, such as *least, best, outline, describe*, and the like. In this way you will tend to answer as requested rather than discover on reviewing your paper that you *listed without describing,* that you selected the *worst* choice rather than the *best* choice, etc.

3. If the examination is of the objective or so-called multiple-choice type, that is, each question will also give a series of possible answers: A, B, C, or D, and you are called upon to select the best answer and write the letter next to that answer on your answer paper, it is advisable to start answering each question in turn. There may be anywhere from 50 to 100 such questions in the three or four hours allotted and you can see how much time would be taken if you read through all the questions before beginning to answer any. Furthermore, if you come across a question or a group of questions which you know would be difficult to answer, it would undoubtedly affect your handling of all the other questions.

4. If the examination is of the essay-type and contains but a few questions, it is a moot point as to whether you should read all the questions before starting to answer any one. Of course if you are given a choice, say five out of seven and the like, then it is essential to read all the questions so you can eliminate the two which are most difficult. If, however, you are asked to answer all the questions, there may be danger in trying to answer the easiest one first because you may find that you will spend too much time on it. The best technique is to answer the first question, then proceed to the second, etc.

5. Time your answers. Before the examination begins, write down the time it started, then add the time allowed for the examination and write down the time it must be completed, then divide the time available somewhat as follows:

   a. If 3 ½ hours are allowed, that would be 210 minutes. If you have 80 objective-type questions, that would be an average of about 2 ½ minutes per question. Allow yourself no more than 2 minutes per question, or a total of 160 minutes, which will permit about 50 minutes to review.

   b. If for the time allotment of 210 minutes, there are 7 essay questions to answer, that would average about 30 minutes a question. Give yourself only 25 minutes per question so that you have about 35 minutes to review.

6. The most important instruction is *to read each question* and make sure you know what is wanted. The second most important instruction is to *time yourself properly* so that you answer every question. The third most important instruction is to *answer every question.* Guess if you have to but include something for each question, Remember that you will receive no credit for a blank and will probably receive some credit if you write something in answer to an essay question. If you guess a letter, say "B" for a multiple-choice question, you may have guessed right. If you leave a blank as the answer to a multiple-choice question, the examiners may respect your feelings but it will not add a point to your score. Some exams may penalize you for wrong answers, so in such cases *only*, you may not want to guess unless you have some basis for your answer.

7. Suggestions

   a. Objective-Type Questions

      (1) Examine the question booklet for proper sequence of pages and questions.

      (2) Read all instructions carefully.

      (3) Skip any question which seems too difficult; return to it after all other questions have been answered.

      (4) Apportion your time properly; do not spend too much time on any single question or group of questions.

      (5) Note and underline key words—*all, most, fewest, least, best, worst, same, opposite.*

      (6) Pay particular attention to negatives.

      (7) Note unusual option, e.g., unduly long, short, complex, different or similar in content to the body of the question.

      (8) Observe the use of "hedging" words—*probably, may, most likely, etc.*

      (9) Make sure that your answer is put next to the same number as the question.

      (10) Do not second guess unless you have good reason to believe the second answer is definitely more correct.

      (11) Cross out original answer if you decide another answer is more accurate; do not erase, *until* you are ready to hand your paper in.

      (12) Answer all questions; guess unless instructed otherwise.

      (13) Leave time for review.

b. Essay-Type Questions

(1) Read each question carefully.

(2) Determine exactly what is wanted. Underline key words or phrases.

(3) Decide on outline or paragraph answer.

(4) Include many different points and elements unless asked to develop any one or two points or elements.

(5) Show impartiality by giving pros and cons unless directed tu select one side only.

(6) Make and write down any assumptions you find necessary to answer the question.

(7) Watch your English, grammar, punctuation, choice of words.

(8) Time your answers; don't crowd material.

8. Answering the Essay Question

Most essay questions can be answered by framing the specific response around several key words or ideas. Here are a few such key words or ideas:

M's: manpower, materials, methods, money, management

P's: purpose, program, policy, plan, procedure, practice, problems, pitfalls, personnel, public relations

a. Six basic steps in handling problems:

(1) preliminary plan and background development

(2) collect information, data and facts

(3) analyze and interpret information, data and facts

(4) analyze and develop solutions as well as make recommendations

(5) prepare report and sell recommendations

(6) install recommendations and follow up effectiveness

b. Pitfalls to Avoid

(1) *Taking Things for Granted*
A statement of the situation does not necessarily imply that each of the elements is necessarily true; for example, a complaint may be invalid and biased so that all that can be taken for granted is that a complaint has been registered

(2) *Considering only one side of a situation*
Wherever possible, indicate several alternatives and then point out the reasons you selected the best one.

(3) *Failing to indicate follow up*
Whenever your answer indicates action on your part, make certain that you will take proper follow-up action to see how successful your recommendations, procedures, or actions turn out to be.

(4) *Taking too long in answering any single question*
Remember to time your answers properly.

EXAMINATION SECTION

EXAMINATION SECTION

TEST 1

DIRECTIONS: Each question or incomplete statement is followed by several suggested answers or completions. Select the one that BEST answers the question or completes the statement. *PRINT THE LETTER OF THE CORRECT ANSWER IN THE SPACE AT THE RIGHT.*

1. There are now ____ states in the United States. 1. A
   A. 50 B. 49 C. 48 D. 51

2. The people who live along our west coast in California are in the ____ Time Zone. 2. C
   A. Mountain B. Central C. Pacific D. Eastern

3. Lava means 3. B
   A. crater B. liquid rock
   C. eruption D. volcano

4. Some of the rocky surfaces in the central part of the United States were caused by 4. C
   A. earthquakes B. volcanoes C. glaciers D. floods

5. The *only* one of the Great Lakes that is completely within the boundaries of the United States is Lake 5. C
   A. Erie B. Huron C. Michigan D. Superior

6. The mountain range which is found in the western part of the United States is called the 6. D
   A. Appalachian B. Blue Ridge
   C. Ozark Plateau D. Rockies

7. The coastal plain region in the United States is found in the 7. B
   A. Mountain States B. Southern States
   C. the North Central States D. Appalachian Mountains

8. One would expect to find the Piedmont Region of the United States near the 8. A
   A. Appalachian Mountains B. Rocky Mountains
   C. Great Plains D. Great Lakes

9. Distances which are measured in degrees are measured north and south from the 9. D
   A. Prime Meridian B. North and South Poles
   C. International Date Line D. Equator

10. The eastern half of the United States was explored MOSTLY by men from 10. D
    A. Germany B. Holland C. Spain D. England

11. One of the LARGEST regions added to the United States was 11. D
A. Hawaii
B. the Gadsden Purchase
C. the Florida Purchase
D. the Louisiana Purchase

Questions 12-14.

DIRECTIONS: Questions 12 to 14 are based on the compass below.

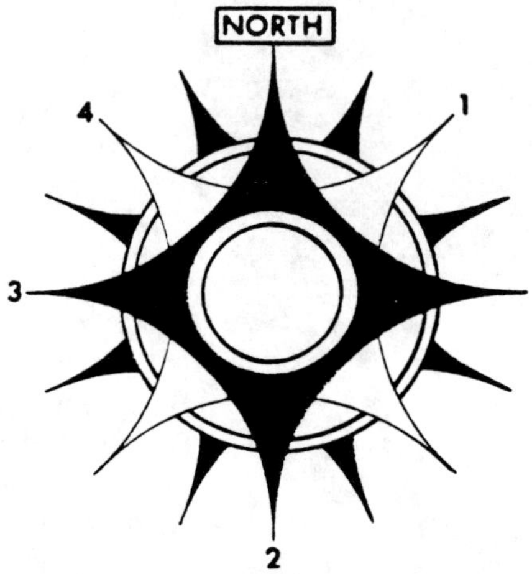

12. Point ____ on the compass above would be to the right of a person facing the sunrise. 12. B
A. 1 B. 2 C. 3 D. 4

13. Point ____ on the compass above is located 45 degrees east of due north. 13. A
A. 1 B. 2 C. 3 D. 4

14. Which direction is indicated by Point 3 on the compass? 14. C
A. East
B. Southwest
C. West
D. Northwest

15. Geography may be defined as the science or study that deals with the 15. C
A. observation and interpretation of the positions, distribution, dimensions, motion, composition, and evolution of celestial bodies and phenomena
B. origin, history, and structure of the earth
C. earth and its features, including the effects of human activity
D. phenomena of the atmosphere, especially weather conditions

16. Geography is *closely* related to the system humans have devised to measure the passage of time because our concept of time is based upon 16. A
A. the movements of the planet upon which we live
B. religious beliefs that originated in different geographic areas
C. clocks and calendars
D. scientific evidence

17. ____ falls within the third century A.D. 17. A
A. A.D. 210 B. 360 B.C. C. A.D. 315 D. A.D. 450

18. A ____ is required for the earth to complete one revolution around the sun. 18. D
A. day B. month C. season D. year

19. A(n) ____ has been developed to fix the divisions of time in seconds, minutes, and hours. 19. D
A. calendar B. almanac
C. astrolabe D. none of the above

20. ____ is an abbreviation used to designate time that has elapsed since the starting point of our calendar. 20. A
A. A.D. B. A.M. C. B.C. D. P.M.

21. A plane left Tokyo early Monday morning, crossed the International Date Line, and landed ten hours later in Los Angeles. 21. A
The plane landed on
A. Sunday B. Monday C. Tuesday D. Wednesday

22. Minor errors in our present calendar are corrected by *leap years*. 22. A
During a leap year,
A. one day is added to the calendar
B. one day is dropped from the calendar
C. two hours are dropped from each month
D. all clocks are set forward one hour

23. A(n) ____ projection is shown on the map projection at the right. 23. A
A. conic
B. Mercator
C. polar
D. interrupted

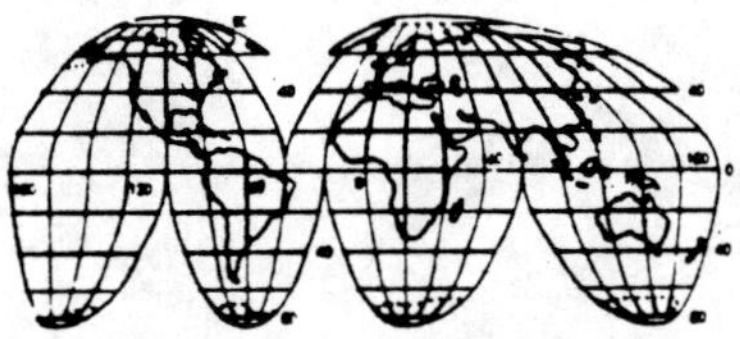

Questions 24-25.

DIRECTIONS: Questions 24 and 25 are based on the map below.

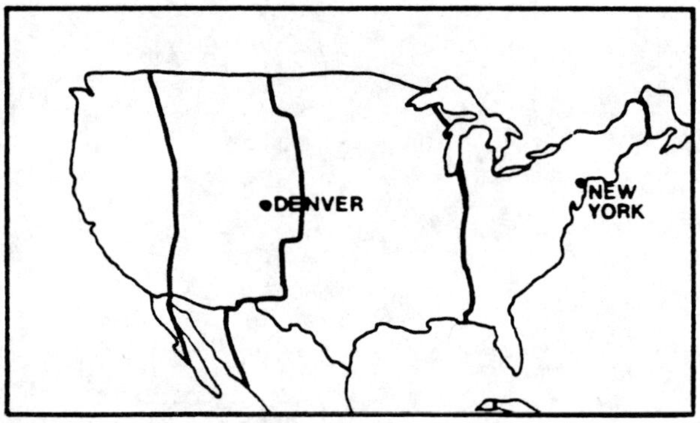

24. If it is 5:00 P.M. in New York, it is ____ P.M. in Denver. 24. A
A. 3:00 B. 4:00 C. 6:00 D. 7:00

25. If it is 10:00 A.M. in St. Louis, Missouri, it is ____ A.M. in San Francisco, California. 25. A
A. 8:00 B. 9:00 C. 10:00 D. 11:00

KEY (CORRECT ANSWERS)

1. A
2. C
3. B
4. C
5. C

6. D
7. B
8. A
9. D
10. D

11. D
12. B
13. A
14. C
15. C

16. A
17. A
18. D
19. D
20. A

21. A
22. A
23. A
24. A
25. A

TEST 2

DIRECTIONS: Each question or incomplete statement is followed by several suggested answers or completions. Select the one that BEST answers the question or completes the statement. *PRINT THE LETTER OF THE CORRECT ANSWER IN THE SPACE AT THE RIGHT.*

1. One of the MOST important minerals found in the British Isles is — 1. C
   A. gold B. diamonds C. coal D. bauxite

2. One of the MOST important sea ports in Great Britain is — 2. B
   A. Le Havre B. Liverpool
   C. Amsterdam D. Rotterdam

3. Linen is made from — 3. B
   A. cotton B. flax C. wool D. rayon

4. If you were to visit ____, you would be able to see the Parliament Building. — 4. A
   A. London B. Glasgow
   C. Southhampton D. Manchester

5. Many fiords are found along the coast of — 5. A
   A. Norway B. France C. Holland D. Poland

6. *Usually,* ____ is referred to as being one of the Scandinavian countries. — 6. A
   A. Sweden B. Poland C. Germany D. Belgium

7. Laplanders are found in northern — 7. A
   A. Finland B. Poland C. Belgium D. France

8. One of the LARGER islands in the North Atlantic is — 8. C
   A. Sardinia B. Cuba C. Iceland D. Samoa

9. One of the low countries of Europe is — 9. C
   A. Austria B. Switzerland
   C. Holland D. Spain

10. Deltas are *usually* found — 10. B
    A. near fiords
    B. at the mouth of large rivers
    C. on a coral reef
    D. near volcanoes

11. Polders are — 11. A
    A. lands that are reclaimed from the sea
    B. lands that are cleared of timber
    C. found in desert areas
    D. jungle areas

12. England and ____ are both governed by a queen. 12. D
  A. France B. Denmark C. Sweden D. Holland

13. Until recently, the Congo region in Africa was owned and governed by 13. D
  A. Holland B. France C. Portugal D. Belgium

14. One of the LARGEST food crops produced in northern Europe is 14. D
  A. rice B. sugar cane C. coffee D. potatoes

15. Textiles are 15. C
  A. food B. a group of minerals
  C. cloth D. none of the above

Questions 16-25.

DIRECTIONS: Use the map below to answer Questions 16 to 25.

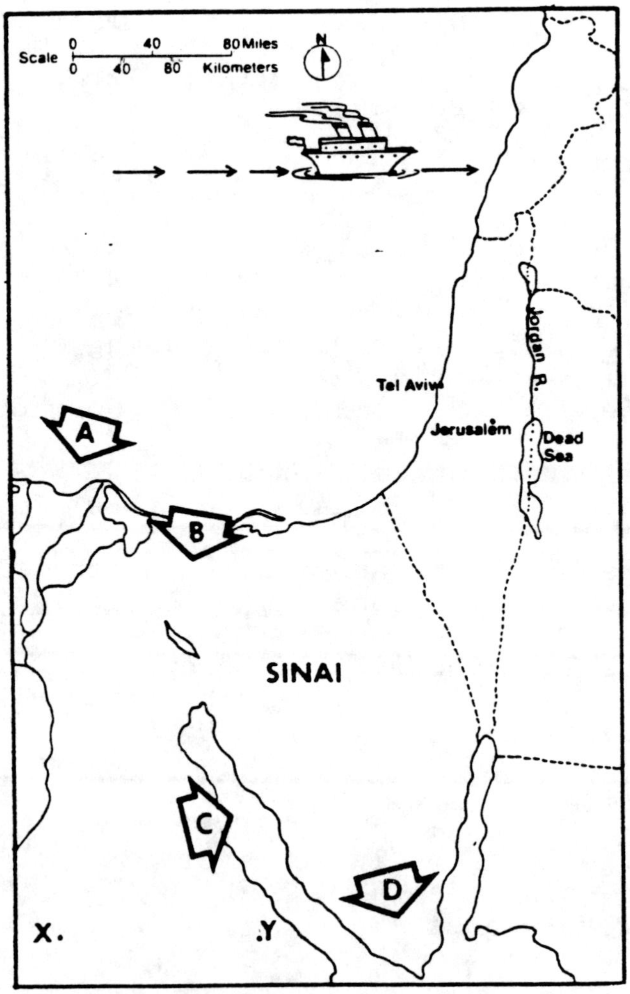

16. Arrow ____ is pointing to an isthmus. 16. B
  A. A B. B C. C D. D

17. Arrow _____ is pointing to a gulf. 17. C
   A. A   B. B   C. C   D. D

18. Arrow _____ is pointing to a delta. 18. A
   A. A   B. B   C. C   D. D

19. Arrow _____ is pointing to a peninsula. 19. D
   A. A   B. B   C. C   D. D

20. Using the straight edge of this paper, the distance between point X and point Y is _____ Km. 20. C
   A. 80   B. 100   C. 160   D. 200

21. Points X and Y are located in 21. C
   A. Asia   B. Egypt   C. Africa   D. India

22. Point X would *most likely* be the site of a(n) 22. B
   A. glacier   B. oasis   C. harbor   D. atoll

23. The map shows a region of the world *generally* referred to as the 23. C
   A. Sahara   B. Far East
   C. Middle East   D. Horn

24. The ship is sailing in the 24. B
   A. Atlantic Ocean   B. Mediterranean Sea
   C. Red Sea   D. Indian Ocean

25. The ship is apparently headed to a port in 25. B
   A. Israel   B. Lebanon
   C. Saudi Arabia   D. Jordan

KEY (CORRECT ANSWERS)

1. C
2. B
3. B
4. A
5. A

6. A
7. A
8. C
9. C
10. B

11. A
12. D
13. D
14. D
15. C

16. B
17. C
18. A
19. D
20. C

21. C
22. B
23. C
24. B
25. B

TEST 3

DIRECTIONS: Each question or incomplete statement is followed by several suggested answers or completions. Select the one that BEST answers the question or completes the statement. *PRINT THE LETTER OF THE CORRECT ANSWER IN THE SPACE AT THE RIGHT.*

1. One of the territories which is owned by the United States is 1. D
   A. Hawaii B. the Philippine Islands
   C. Cuba D. Puerto Rico

2. One can expect to find desert regions in the ____ region. 2. A
   A. southwestern B. northeastern
   C. Great Lakes D. southeastern

3. The great seaport in the United States is 3. A
   A. New York B. Chicago
   C. Washington, D.C. D. Norfolk

4. The *extreme* northeastern part of the United States is referred to as 4. A
   A. New England B. the lake region
   C. the great farm belt D. the desert region

5. One of the states that is *usually* included in the New England states is 5. A
   A. Massachusetts B. Ohio
   C. Kansas D. Florida

6. One of the MOST important crops produced in the New England states is 6. A
   A. potatoes B. cotton
   C. sugar cane D. citrus fruits

7. Much of the money which a New Englander makes comes from 7. D
   A. mining B. cotton
   C. fur trapping D. fishing

8. Leather is made from 8. C
   A. wood B. cloth C. hides D. plastic

9. MOST of the paper made in the United States is made from 9. A
   A. wood B. rags
   C. wool D. none of the above

10. One of the MOST important cities in the New England states is 10. B
    A. New Orleans B. Boston
    C. Philadelphia D. Richmond

11. One of the states in the Middle Atlantic states is — 11. A
   A. New Jersey B. Virginia
   C. West Virginia D. South Carolina

12. The wide mouth of a river in which the tide of the sea rises and falls as it does in the sea itself is called a(n) — 12. B
   A. harbor B. estuary
   C. sea wall D. lock

13. One of the LARGEST cities in the Middle Atlantic states is — 13. D
   A. New Orleans B. Cleveland
   C. Chicago D. Philadelphia

14. The Erie Canal or New York State Barge Canal connects Lake Erie with the ____ River. — 14. D
   A. Delaware B. Potomac
   C. Connecticut D. Hudson

15. One of our LARGEST coal-producing states is — 15. D
   A. New York B. Maryland
   C. New Jersey D. Pennsylvania

16. Suburbs are — 16. C
   A. farming areas
   B. mining areas
   C. nearby residential towns of large cities
   D. bomb shelters

17. New York State is known as the ____ State. — 17. D
   A. Keystone B. Garden C. Diamond D. Empire

18. One would expect to find the tundra region of Canada in the ____ part. — 18. A
   A. northern B. southern C. eastern D. western

19. The LARGEST and MOST important river in Canada is the — 19. D
   A. Hudson River B. Yukon River
   C. Peace River D. St. Lawrence River

Questions 20-25.

DIRECTIONS: Questions 20 to 25 are based on the following diagrams.

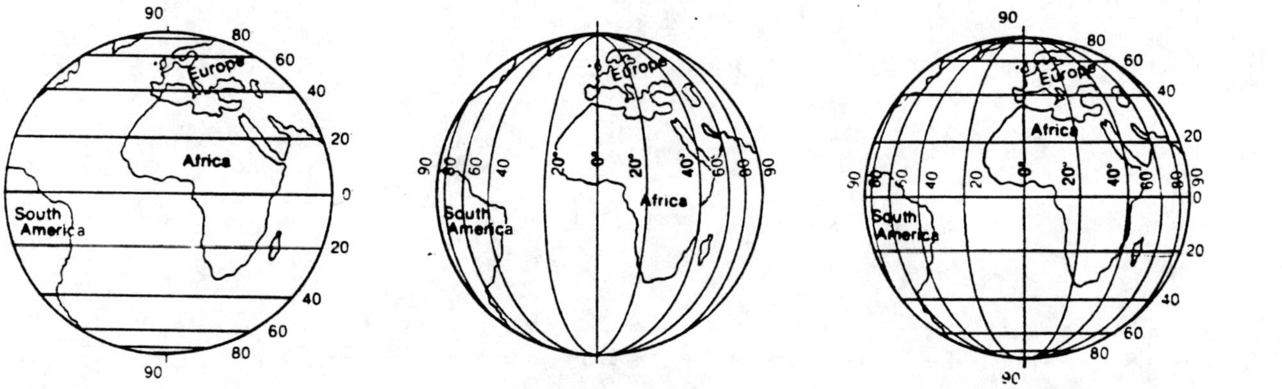

1 2 3

20. ____ are shown in Diagram 1. 20. A
A. Lines of latitude B. Lines of longitude
C. Great circles D. Hemispheres

21. Which statement is FALSE about lines of latitude? 21. C
A. These lines run parallel, and for this reason they are often called *parallels*.
B. Each degree of latitude covers approximately the same number of kilometers on the surface of the earth.
C. They measure distance east and west of the equator.
D. The equator is one of them.

22. It is FALSE that lines of longitude 22. B
A. run north and south and are used to measure distance east and west
B. are numbered east and west of a line at 0°, called the equator
C. are sometimes referred to as *meridians*
D. are numbered east and west of a line at zero meridian (0°), which is called the prime meridian

23. 40° ____ passes through the Italian peninsula. 23. B
A. North Longitude B. North Latitude
C. East Longitude D. West Latitude

24. In Diagram 3, which set of grid coordinates is located on the continent of South America? 24. D
A. 20° North Latitude, 40° West Longitude
B. 0° Latitude, 20° East Longitude
C. 20° West Latitude, 40° South Longitude
D. 20° South Latitude, 60° West Longitude

25. In Diagram 3, grid coordinates 20° South Latitude, 20° East Longitude are located in 25. C
A. North America B. South America
C. Africa D. Europe

KEY (CORRECT ANSWERS)

1. D
2. A
3. A
4. A
5. A
6. A
7. D
8. C
9. A
10. B
11. A
12. B
13. D
14. D
15. D
16. C
17. D
18. A
19. D
20. A
21. C
22. B
23. B
24. D
25. C

TEST 4

DIRECTIONS: Each question or incomplete statement is followed by several suggested answers or completions. Select the one that BEST answers the question or completes the statement. *PRINT THE LETTER OF THE CORRECT ANSWER IN THE SPACE AT THE RIGHT.*

1. MOST of the French Canadians live in — 1. D
   A. Ontario B. Alberta
   C. British Columbia D. Quebec

2. One of the Atlantic Provinces in Canada is — 2. A
   A. Nova Scotia B. British Columbia
   C. Alberta D. Saskatchewan

3. One would use a trawler if he were a — 3. A
   A. fisherman B. miner
   C. farmer D. civil servant

4. One of the St. Lawrence Valley Provinces in Canada is — 4. D
   A. Nova Scotia B. New Brunswick
   C. Prince Edward Island D. Quebec

5. The capital of Canada is — 5. C
   A. Quebec B. Toronto C. Ottawa D. Hamilton

6. Cellulose comes from — 6. A
   A. wood B. cotton C. fish D. mines

7. One of the famous canals in Canada is the — 7. B
   A. New York State Barge B. Welland
   C. Panama D. Suez

8. One of the HIGHEST ranges of mountains in Europe is the — 8. C
   A. Rockies B. Andes C. Alps D. Atlas

9. One of the navigable rivers in Europe is the — 9. C
   A. Amazon B. Tigris C. Danube D. Ohio

10. Paris is the capital of — 10. D
    A. Germany B. Denmark C. Holland D. France

11. The two MAIN islands of the British Isles are Great Britain and — 11. A
    A. Ireland B. Jersey C. Sicily D. Sardinia

12. One of the *present* dominions of the British Commonwealth of Nations is — 12. C
    A. Cuba B. Ireland
    C. Australia D. British Honduras

13. One of the LARGEST port cities in southern France is 13. B
A. Bordeaux B. Marseille C. Rome D. Bern

14. One country in Europe that is divided as a direct result of World War II is 14. A
A. Germany B. Hungary C. Poland D. Italy

15. The MOST important ocean trade route in the world is from 15. A
A. western Europe to the United States
B. the United States to South America
C. Africa to South America
D. Africa to Australia

16. The diagram at the right shows the position of the earth when it is 16. C
A. winter in the northern hemisphere
B. summer in the southern hemisphere
C. daylight in the western United States
D. all of the above

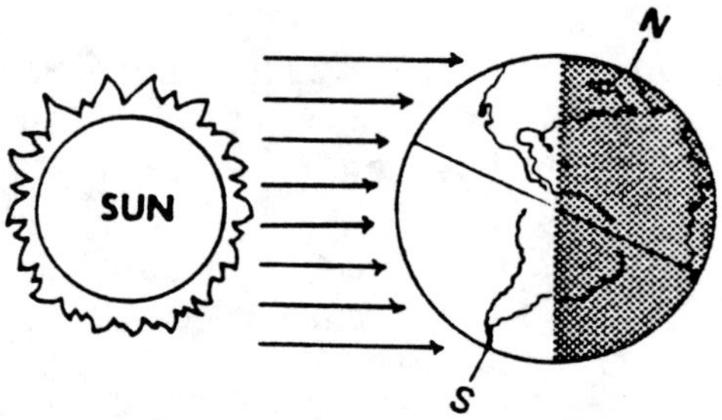

17. In the diagram shown in the previous question, *approximately* what time is it in New York City? 17. C
A. 4:00 - 6:00 A.M. B. 8:00 - 10:00 A.M.
C. 4:00 - 6:00 P.M. D. 10:00 - 12:00 P.M.

18. 18. D

WORLD AIR DISTANCE TABLE

	Azores Islands	Berlin	Buenos Aires	Calcutta	Cape Town	Chicago	
Azores Islands		2185	5417	6549	5854	3093	10209
Berlin, Germany	2185		7376	4376	5977	4402	8036
Buenos Aires, Argentina	5417	7376		10242	4270	5596	9127
Calcutta, India	6549	4376	10242		6026	7981	3744
Cape Town, South Africa	5854	5977	4270	6026		8449	6947
Chicago, U.S.A.	3093	4402	5596	7981	8449		9346
Darwin, Australia	10209	8036	9127	3744	6947	9346	

What is the air distance between Buenos Aires and Chicago?
A. 5417 miles B. 3093 miles
C. 9127 miles D. 5596 miles

19. Which point is TRUE about all calendars currently used in different regions of the world? 19. B
A. They are complex and difficult for people to understand.
B. Each is based upon a commonly-accepted starting point in history.
C. They all begin with the birth or death of some famous person.
D. All were developed for religious purposes.

20. Which figure below is an outline map of the State of Louisiana? 20. A

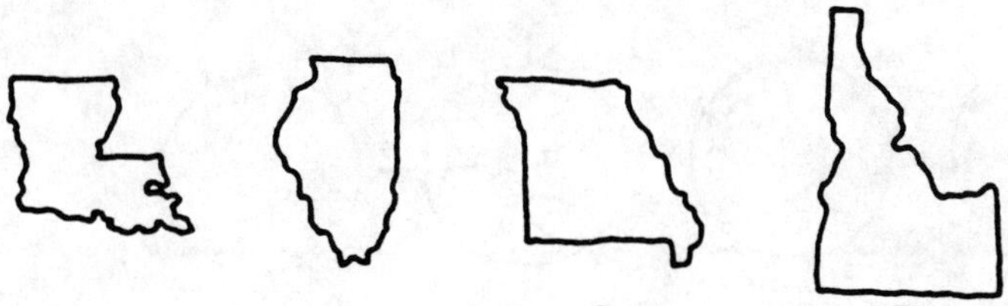

A. B. C. D.

21. The *approximate* elevation at point X on the map shown at the right is ____ meters. 21. B

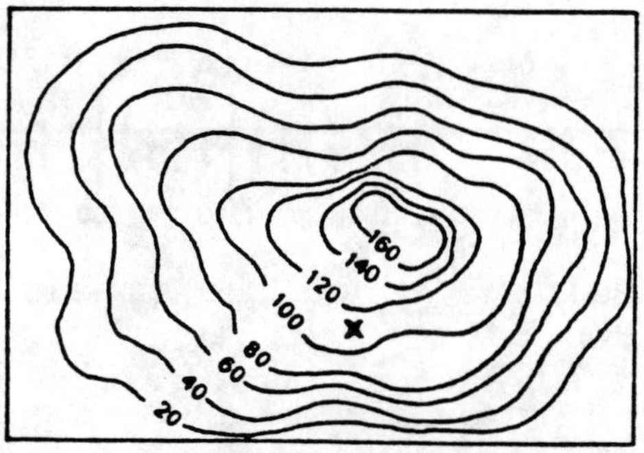

A. 80
B. 100
C. 120
D. 140

22. A ____ map is shown at the right. 22. A

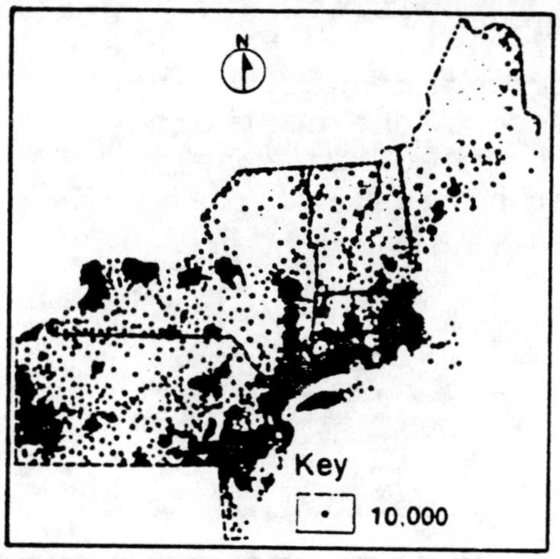

A. population density
B. weather
C. contour
D. climate

23. ____ would be natural vegetation you would expect to find in a savanna region. 23. C

A. Mosses and stunted arctic plants
B. Needleleaf evergreen and deciduous forests
C. Tropical grass with scattered deciduous woodlands
D. Tropical broadleaf evergreen forests

24. ____ would be natural vegetation you would expect to find in a tundra region. 24. A

A. Mosses and stunted arctic plants
B. Needleleaf evergreen and deciduous forests
C. Tropical grass with scattered deciduous woodlands
D. Tropical broadleaf evergreen forests

25. Which figure below shows the CORRECT axis and rotational direction of the earth? 25. B

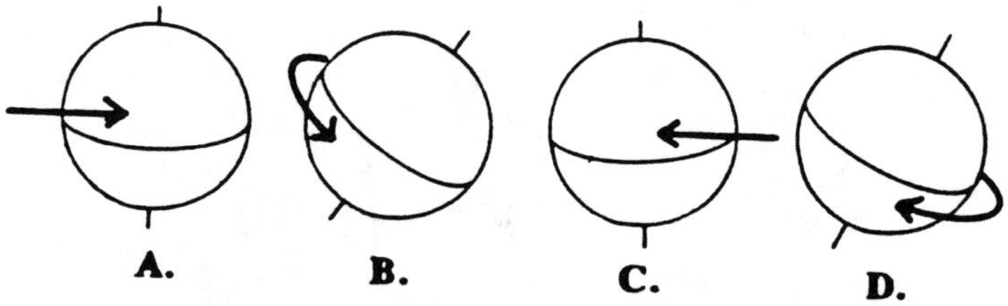

26. 26. A

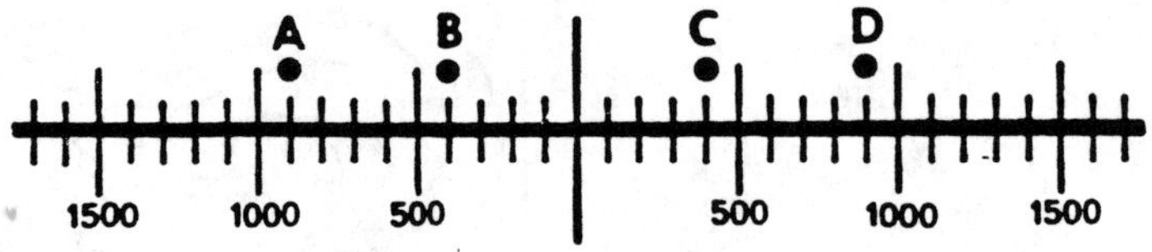

Point ____ on the timeline above indicates the year 900 B.C.

A. A B. B C. C D. D

27. Angle ____ in the drawing at the right shows the angle of sunlight that strikes the surface of the northern hemisphere during the summer months. 27. D

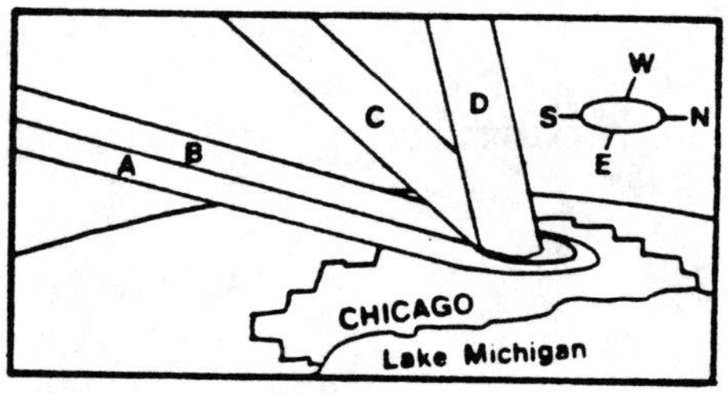

A. A
B. B
C. C
D. D

Questions 28-30.

DIRECTIONS: Questions 28 to 30 are based on the following map.

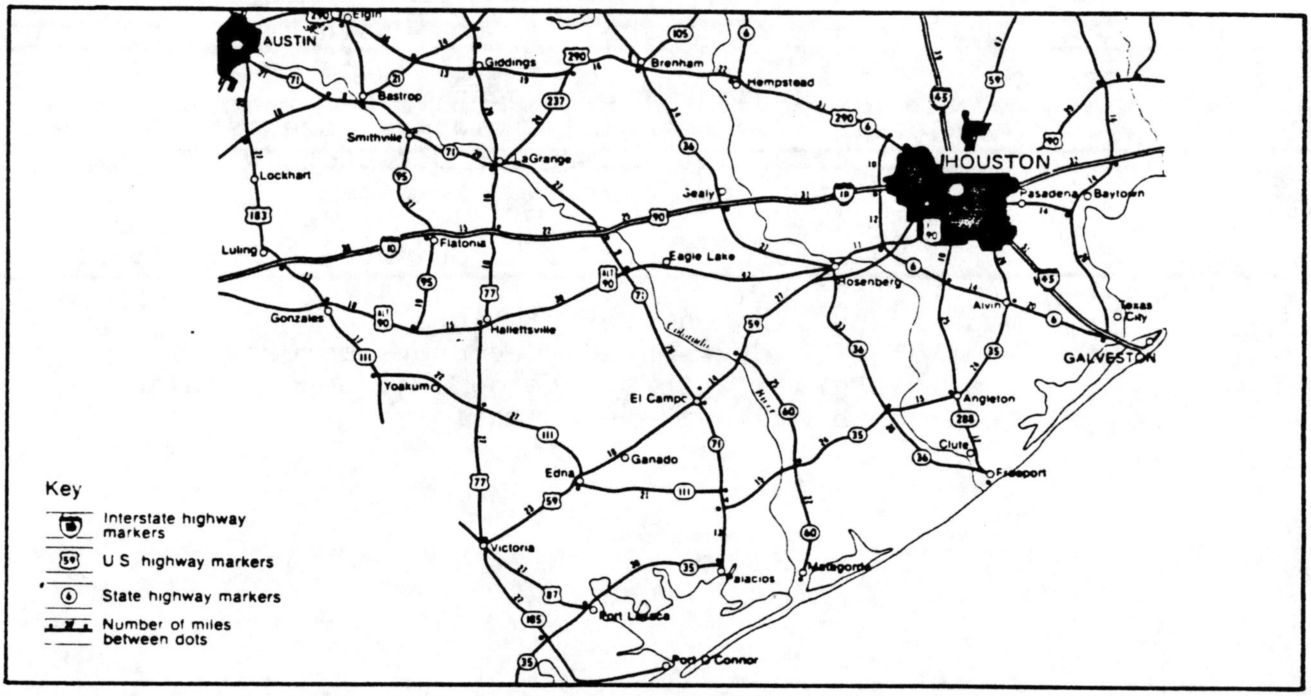

28. On the road map shown above, the SHORTEST route between the towns of Flatonia and Gonzales is 28. A
   A. State 95 to Alternate U.S. 90
   B. Interstate 10 to U.S. 183
   C. Interstate 10 to U.S. 77 to Alternate U.S. 90
   D. State 95 to Alternate U.S. 90 to U.S. 77

29. ____ would *most likely* be called the *Scenic River Route* according to the map. 29. C
   A. Interstate 10
   B. U.S. 59
   C. State 71
   D. Alternate U.S. 90

30. The MOST direct route from Eagle Lake to the Texas state capital is 30. C
   A. Alternate U.S. 90 to Interstate 10
   B. Alternate U.S. 90
   C. Alternate U.S. 90 to State 71
   D. not indicated on this map

KEY (CORRECT ANSWERS)

1. D	11. A	21. B
2. A	12. C	22. A
3. A	13. B	23. C
4. D	14. A	24. A
5. C	15. A	25. B
6. A	16. C	26. A
7. B	17. C	27. D
8. C	18. D	28. A
9. C	19. B	29. C
10. D	20. A	30. C

EXAMINATION SECTION

TEST 1

DIRECTIONS: Each question or incomplete statement is followed by several suggested answers or completions. Select the one that BEST answers the question or completes the statement. *PRINT THE LETTER OF THE CORRECT ANSWER IN THE SPACE AT THE RIGHT.*

1. Which one of the following CORRECTLY pairs a French city with the river on which it is located? 1. D
   A. Lyons - Garonne
   B. Rouen - Rhone
   C. Toulouse - Seine
   D. Tours - Loire

2. Which one of the following is TRUE of continental shelves? 2. D
   A. The rocky bottoms are inhospitable to sea life and fish.
   B. Geologically, they are distinctly separated from the adjoining continents.
   C. Because of their depth, geologists know little about the shelves.
   D. A nation's jurisdiction extends to its continental shelf.

3. Which of the following are CORRECTLY paired? 3. C
   I. Gobi Desert in Mongolian Asia
   II. Sahara Desert in North Africa
   III. Kalahari Desert in Central Australia
   IV. Atacama Desert in Chile

   The CORRECT answer is:
   A. I, II, III
   B. III, IV
   C. I, II, IV
   D. I, II, III, IV

4. Which one of the following Asian nations is a *food-surplus* rather than a *food-deficient* region? 4. A
   A. Burma (Myanmar)
   B. India
   C. Malaya
   D. North Korea

5. The areas of extreme temperatures (hot and cold) are located in the 5. C
   A. equatorial region
   B. frigid zones
   C. temperate zones
   D. torrid zone

6. In the vicinity of the 100th meridian in the United States, a farmer would MOST likely 6. D
   A. grow citrus fruits
   B. specialize in dairy farming
   C. cultivate a *truck* farm
   D. grow wheat with *dry farming* methods

7. Which one of the following regarding the physical conditions of Africa south of the Sahara is INCORRECT? 7. C
   A. It is essentially a great plateau with fringing mountains along parts of the edge.
   B. Most of the rivers are broken by falls and rapids close to the coast.
   C. Coastal plains occupy a large portion of the total area.
   D. The coast is critically lacking in good harbors.

8. Which one of the following statements is NOT true about the Po Valley? 8. B
   A. It is the most densely populated region of Italy.
   B. It has extensive coal reserves.
   C. Its alluvial soil is accompanied by adequate rainfall and mild temperatures.
   D. The conditions are favorable for the development of water power.

9. The United States must import MOST of its supply of 9. C
   A. copper and iron
   B. lead and zinc
   C. manganese and nickel
   D. phosphates and sulphur

10. Which of the following land areas would be shown MOST accurately on the mercator projection map? 10. C
    A. Greenland - Iceland
    B. Alaska - Eastern Siberia
    C. Colombia - Venezuela
    D. Northern Canada - Northern Scandinavia

11. Winds that reverse directions in the summer and winter are the 11. A
    A. monsoons
    B. siroccos
    C. typhoons
    D. trade winds

12. The Volga River empties into the 12. A
    A. Caspian Sea
    B. Black Sea
    C. White Sea
    D. Don River

13. A population distribution map of South America would show that 13. B
    A. the population is spread evenly throughout the continent
    B. 75% of the people live within 200 miles of the coast
    C. the majority of the people live in the Amazon basin
    D. the majority of the people live in the mountain plateau

14. The countries of South America buy the MAJOR portion of their imports from 14. C
    A. other South American countries
    B. England
    C. United States
    D. Japan

15. In which of the following pairs of countries is the average per capita income LOWEST? 15. B
    A. Peru and Venezuela
    B. Bolivia and Haiti
    C. Cuba and Costa Rica
    D. Uruguay and Argentina

16. Which one of the following ocean currents touches the southwest coast of Africa? 16. A
    A. Benguela
    B. Gulf Stream
    C. Humboldt
    D. Kuroshio

17. The two MOST populated African nations are 17. B
    A. Algeria and Sudan
    B. Ethiopia and Egypt
    C. Morocco and Kenya
    D. Ghana and Union of South Africa

18. It is situated along the Gulf of Guinea. Cocoa, its chief export, constitutes one-third of the world's supply. Its future industrialization hinges on its Volta River Project. It gained its independence from Great Britain in 1957. This is a description of 18. B
    A. Nigeria
    B. Ghana
    C. Gabon
    D. Ivory Coast

19. Which one of the following mineral products is NOT an important export from South American countries? 19. B
    A. Bauxite B. Coal C. Copper D. Tin

20. Mistral, bora, sirocco are 20. D
    A. glacial formations found in the Swiss Alpine and the Carpathian ranges
    B. tropical typhoons that originate in the Indian and Pacific Oceans
    C. rock formations found in the Italian Dolomites and French Massif Central
    D. winds to be found in Mediterranean and North African areas

21. Which one of the following was the FIRST Latin-American country in which the government confiscated large landed estates for distribution to peasants? 21. A
    A. Mexico B. Cuba C. Bolivia D. Guatemala

22. Which one of the following areas of the United States has a Mediterranean-type climate? 22. B
    A. Along the Gulf of Mexico
    B. California
    C. Arizona
    D. Florida's east coast, Key West to Miami

23. Alluvial fans are land formations resulting from 23. B
    A. an erosion process
    B. a deposition process
    C. glacial movement
    D. a weathering process

24. The map projection which shows the meridians of longitude as parallel lines is the 24.A
   A. Mercator B. Lambert Conformal
   C. Polyconic D. Homolographic

25. Which climatic impact does the Humboldt Current have? It _____ coast. 25.C
   A. warms the Argentine
   B. cools the Southwest African
   C. cools the Peruvian
   D. warms the Japanese

26. Which of the following had its land area separated by another nation until the establishment of Bangladesh? 26.D
   A. Canada B. Paraguay C. Burma D. Pakistan

27. Which one of the following African nations is landlocked? 27.C
   A. Tanzania B. Kenya C. Chad D. Sudan

28. Which is a characteristic of the northern tundra lands? 28.B
   A. Most of them are located south of the Arctic Circle.
   B. Winter temperatures generally are not as cold as those of the forest belts to the south.
   C. They are generally an area of heavy snowfall.
   D. They are devoid of land animals and land plants.

29. An example of the influence of geography on the history of China is 29.A
   A. its long period of isolation from the western world because of natural barriers
   B. the evenness of population distribution because of its abundance of fertile lands
   C. the early establishment of industries because of its vast natural resources
   D. the importation of religious beliefs from Japan because of the latter's proximity

30. Which one of the following geographic conditions is ideal for the growing of coffee? 30.D
   A. Annual rainfall under 40 inches
   B. Daytime temperature of no more than 80°
   C. Sandy soil
   D. Altitude of 1,000 to 6,000 feet

31. MOST of the land area in Africa south of the Equator consists of 31.D
   A. river valleys B. rain forests
   C. desert D. grasslands

32. The GREAT economic potential of Guinea stems from the fact that it 32.B
   A. is the world's largest exporter of cocoa
   B. possesses large reserves of bauxite
   C. remained part of France when other west African states broke away from France
   D. is one of the few nations in Africa with a multi-party system

33. Which of the following statements about the rivers of South America is TRUE? 33.___
   A. Rivers flowing to the Pacific are long and navigable to the source.
   B. The Orinoco has two thousand miles of unimpeded navigable water through rich industrial areas.
   C. The Amazon system offers twenty thousand miles of navigable waters.
   D. The Parana-La Plata is navigable only at the Atlantic estuary.

34. Which of the following countries share a disputed border area? 34.___
   A. Afghanistan - Pakistan   B. Iraq - Syria
   C. Iran - Syria   D. Turkey - Azerbaijan

35. Austria, Hungary, the Czech Republic and Slovakia are similar in that they all 35.___
   A. have no coastlines
   B. are contiguous to Russia
   C. are members of the Warsaw Pact
   D. have Communist governments

Questions 36-38.

DIRECTIONS: Questions 36 through 38 are to be answered on the basis of the interpretation of the climatic station drawn below.

CITY X

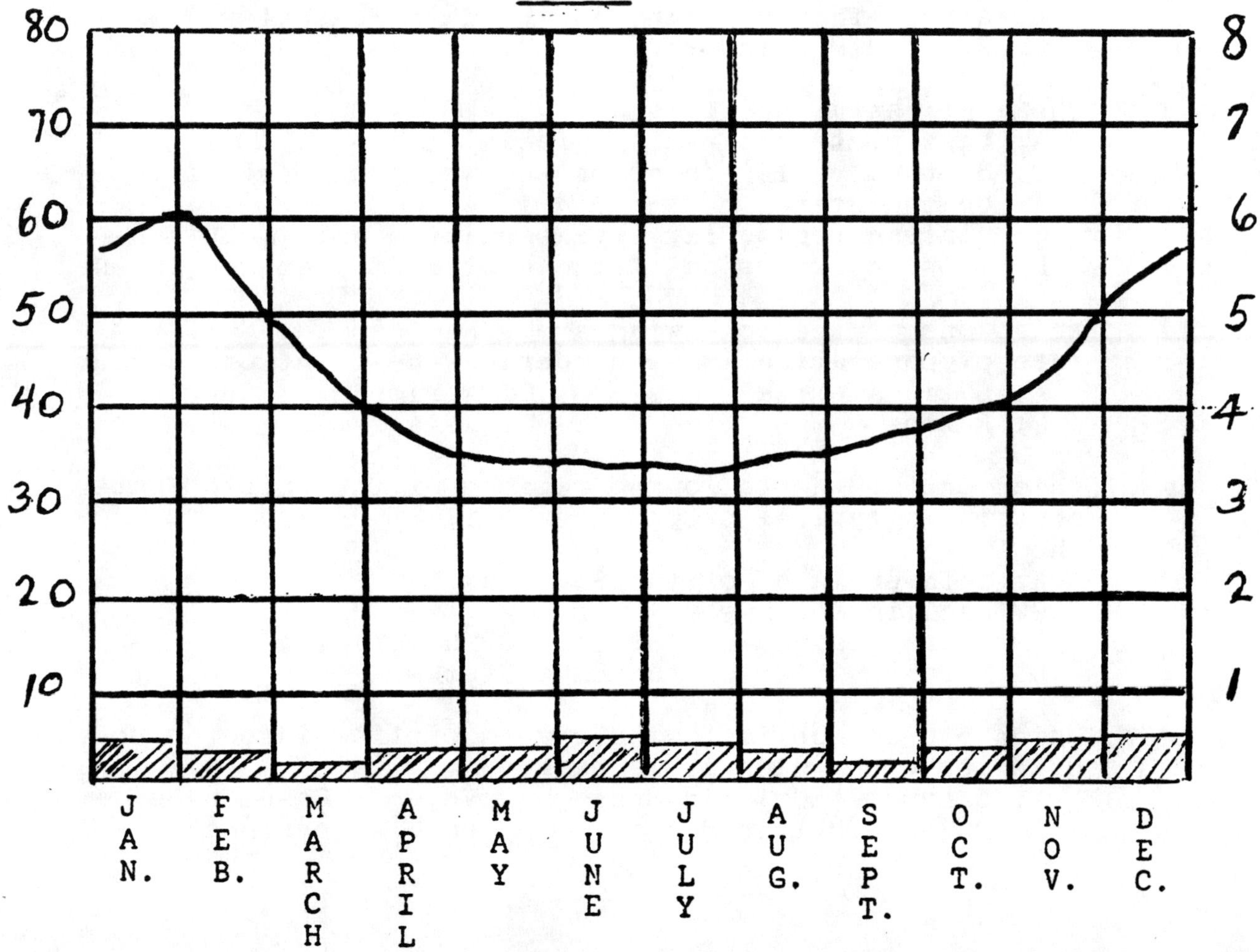

LEGEND

1. The left vertical column indicates temperature in degrees Fahrenheit.
2. The right vertical column indicates precipitation in inches.
3. The curved line indicates the degrees.
4. The boxes with diagonal lines at the base of the chart indicate precipitation.

36. City X is MOST probably located in the 36. B
    A. Tropics
    B. Southern Hemisphere - Middle Latitudes
    C. Northern Hemisphere - Temperate Zone
    D. Sub-Arctic

37. The climate at City X is BEST described as 37. D
    A. Monsoonal B. Tundra
    C. Low latitude steppe D. Middle latitude desert

38. The occupation MOST probably found in the area surrounding City X is 38. A
    A. pastoral nomadic B. rice cultivation
    C. plantation agriculture D. wheat farming

39. Which one of the following twentieth century geographers offered the theory that a nation must expand territorially or decline? 39. B
    A. Alfred Thayer Mahan B. Karl Haushofer
    C. Halford Mackinder D. Ellsworth Huntington

40. The continental shelf is a 40. D
    A. vast plateau of solid glacial ice
    B. drowned valley found on the western coast of the United States
    C. plateau region largely associated with mountains
    D. sloping extension of the continent into shallow seas

41. Which one of the following areas has had the MOST rapid rate of population increase during the twentieth century? 41. A
    A. Latin America B. Africa
    C. Europe D. Asia

42. Which one of the following regions of the United States contains a typical steppe area? 42. C
    The
    A. cutover lands of the Southeast
    B. Pacific Northwest
    C. Great Plains west of the 100th meridian
    D. Northern tier of the Lake States

43. Which one of the following rivers empties into the body of water with which it is paired? 43. B
    A. Danube - Adriatic Sea B. Volga - Caspian Sea
    C. Rhine - Baltic Sea D. Po - Aegean Sea

44. In which one of the following do both countries border on Vietnam? 44. C
   A. Burma and Cambodia B. Thailand and Burma (Myanmar)
   C. Cambodia and Laos D. China and Malaya

45. Geographers refer to the areas in the low latitudes where scattered trees and grasslands are intermingled as 45. B
   A. prairie lands B. savannas
   C. pampas D. tundra regions

46. Which one of the following describes a type of soil? 46. D
   A. Foehn B. Sirocco C. Chinook D. Chernozem

47. A MAJOR ocean current that tends to cause desert conditions in adjacent land areas is the 47. A
   A. Humboldt Current B. Gulf Stream
   C. Japanese Current D. Labrador Current

48. The climatic characteristics of Washington, D.C. are APPROXIMATELY the same as those of 48. A
   A. Tokyo B. Moscow
   C. Los Angeles D. Paris

49. Which one of the following is the CORRECT relationship between density of population and agriculture? 49. A
   A. There is no universal correlation between the relative importance of agriculture and density of population.
   B. The countries which are dominantly agricultural are also the more thinly settled lands.
   C. The agricultural nations of the world are uniformly also the most densely settled areas.
   D. The countries that have witnessed the strongest movements of population from farm to city have suffered the greatest loss in agricultural production.

50. Expressed in statute miles, a degree of longitude 50. A
   A. varies with the parallel of latitude along which it is measured
   B. varies only at the poles
   C. is constant at all latitudes
   D. is equal to a degree of latitude

KEY (CORRECT ANSWERS)

1. D	11. A	21. A	31. D	41. A
2. D	12. A	22. B	32. B	42. C
3. C	13. B	23. B	33. C	43. B
4. A	14. C	24. A	34. A	44. C
5. C	15. B	25. C	35. A	45. B
6. D	16. A	26. D	36. B	46. D
7. C	17. B	27. C	37. D	47. A
8. B	18. B	28. B	38. A	48. A
9. C	19. B	29. A	39. B	49. A
10. C	20. D	30. D	40. D	50. A

TEST 2

DIRECTIONS: Each question or incomplete statement is followed by several suggested answers or completions. Select the one that BEST answers the question or completes the statement. *PRINT THE LETTER OF THE CORRECT ANSWER IN THE SPACE AT THE RIGHT.*

1. Which one of the following types of climate in regions along margins of continents has the following characteristics: average annual rainfall generally below 30 inches, about six to eight months with average temperatures below 65°F, mild climate, dry summer, and rainy winter? 1. B
   A. East coast continental
   B. Dry subtropical-Mediterranean
   C. West coast marine
   D. Humid continental

2. Which one of the following national parks is INCORRECTLY paired with the state in which it is located? 2. A
   A. Acadia - California
   B. Grand Teton - Wyoming
   C. Glacier - Montana
   D. Zion - Utah

3. The length of a degree of longitude at different latitudes from the equator to the North Pole 3. C
   A. remains the same
   B. is not related to the projection
   C. decreases
   D. increases

4. The country with the LARGEST wood pulp production is 4. D
   A. Sweden
   B. Canada
   C. Russia
   D. United States

5. Baku, Grozny, Maracaibo, and Ploesti are areas associated with deposits of 5. A
   A. petroleum
   B. iron ore
   C. uranium
   D. copper

Questions 6-10.

DIRECTIONS: Questions 6 through 10 are to be answered on the basis of the map below.

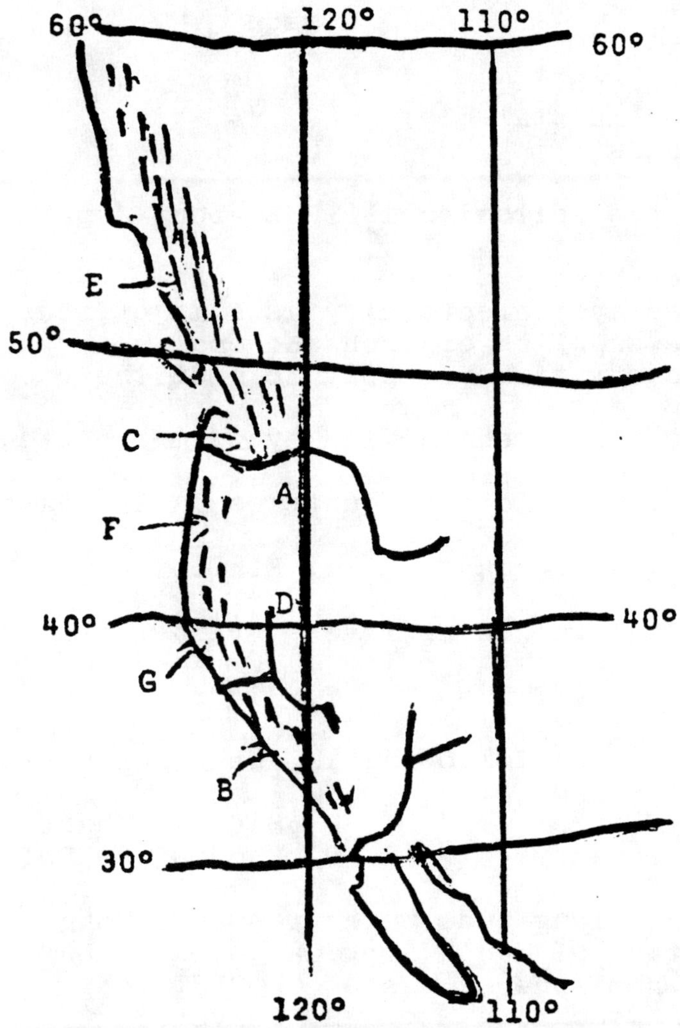

6. What time would it be in a city located at point A if it were 12 Noon at Greenwich? 6. A

   A. 4 A.M. B. 4 P.M. C. 6 A.M. D. 8 P.M.

7. Of the following, the area which would get the LEAST amount of rainfall is 7. D

   A. C B. F C. E D. A

8. An observer will see the sun at point D on June 21 8. A

   A. $16\frac{1}{2}$° south of his zenith
   B. directly overhead
   C. 25° north of his zenith
   D. 30° east of his zenith

9. The winds which directly affect the area between 30° and 60° are the 9. B
A. Northeast Trades
B. Prevailing Westerlies
C. Doldrums
D. Southeast Trades

10. Which one of the following areas is CORRECTLY matched with its climate? 10. B

	Climate	Region
A.	Rainforest	A
B.	Mediterranean	B
C.	Humid Continental	C
D.	Dry Savanna	D

11. Which one of the following is TRUE about Japan and Great Britain? 11. D
Both countries have
A. developed hydroelectricity as a major source of power
B. adequate supplies of iron and coal
C. a sufficient timber supply to meet their domestic needs
D. climates that are modified by warm ocean currents

12. Which one of the following countries in Africa has the LARGEST population? 12. A
A. Nigeria B. Sudan C. Ethiopia D. Kenya

13. Which one of the following rivers drains into the Indian Ocean? 13. D
A. Senegal B. Congo C. Niger D. Zambezi

14. In which one of the following is the country INCORRECTLY paired with its capital? 14. C
A. Kenya - Nairobi
B. Sudan - Khartoum
C. Ghana - Conakry
D. Rhodesia - Salisbury

15. The meridian of longitude which passes through the most westerly portion of South America, when extended into the Northern Hemisphere, passes CLOSEST to 15. D
A. San Francisco
B. Denver
C. Chicago
D. Boston

16. Mexico City is in the same latitude as 16. A
A. Bombay B. Rome C. Osaka D. Alexandria

17. Petroleum is the CHIEF mineral resource in 17. A
A. Romania
B. Albania
C. Bulgaria
D. Serbia

18. Which one of the following statements is NOT true regarding Indonesia? 18. A
A. The population is evenly spread among the larger islands.
B. The leading exports are rubber, petroleum, and tin.
C. Most of the nation lies in the tropical rainforest climatic belt.
D. The islands extend for more than 3,000 miles.

19. Which one of the following is NOT true of Africa? It 19. C
    A. is a vast plateau except for the Atlas mountains and some ranges in the east
    B. is based on a foundation of igneous and sedimentary rocks
    C. is covered almost entirely with steeply folded mountain ranges
    D. lacks natural harbors especially on the west coast

20. The Danube River runs through all of the following countries EXCEPT 20. A
    A. Poland and Albania
    B. Germany and Croatia
    C. Hungary and Austria
    D. Bulgaria and Romania

21. Which one of the following is TRUE of the Amazon Basin? It 21. B
    A. is the third largest producer of natural rubber after Malaya and Indonesia
    B. contains the most extensive single area of equatorial rainforest in the world
    C. provides the coal needed to make use of the iron ore deposits of Minas Geraes
    D. is a major supplier of hardwoods, resins, and gums to world trade

22. Which one of the following continents has no area in which the average annual rainfall exceeds 80 inches? 22. D
    A. Africa
    B. Asia
    C. South America
    D. Australia

23. The valley of which one of the following rivers is the LEAST densely populated? 23. B
    A. Rhine
    B. Amazon
    C. Yangtse
    D. Mississippi

24. Which one of the following cities has a climate that is different from that of Los Angeles? 24. D
    A. Valparaiso
    B. Cape Town
    C. Naples
    D. Canton

25. Which one of the following is NOT correctly associated with a major product? 25. D
    A. Egypt - cotton
    B. Ghana - cacao
    C. Congo - copper
    D. Senegal - petroleum

26. Which one of the following is INCORRECTLY matched? 26. B
    A. India - Deccan Plateau
    B. Pakistan - Plateau of Anatolia
    C. Afghanistan - Hindu Kush Mountains
    D. Iran - Elburz Mountains

27. Arab shows sail from ports in southern Arabia to the east coast of Africa propelled by the 27.A

A. Monsoons
B. Doldrums
C. Westerlies
D. Northeast Trades

28. Which one of the following is INCORRECTLY matched? 28.D

A. Czech Republic - Sudeten
B. Poland - Silesia
C. Romania - Transylvania
D. Bulgaria - Bessarabia

29. Which one of the following about Poland is INCORRECT? 29.A

A. Most of the country is made up of hills.
B. It is transversed by the Oder and Vistula Rivers.
C. It has the seaports of Gydnia and Stettin.
D. It has important deposits of coal.

30. MOST of the world's supply of jute is produced in the 30.D

A. Tigris-Euphrates Plain
B. Amazon Valley
C. Congo Basin
D. Ganges-Brahmaputra Delta Region

31. Which one of the following is MOST NEARLY self-sufficient in agriculture? 31.C

A. England
B. Germany
C. France
D. Belgium

32. Which one of the following is TRUE of the rivers of north-west, continental Europe? 32.A

A. Navigable rivers are ensured by the mild, maritime climate.
B. The rivers freeze solid from late fall to late spring.
C. The rivers flood in the spring and dry out in the summer because of uneven distribution of rainfall.
D. Navigation is made difficult by swift mountain streams.

33. The 74th meridian, which passes through New York, also passes through the 33.B

A. Isthmus of Panama and the Pacific Ocean
B. Peruvian Andes
C. Amazon Basin
D. Brazilian highlands

34. A ship going from London, Eng. to St.Petersburg would pass through all of the following bodies of water EXCEPT the 34.D

A. North Sea
B. Skagerrak-Kattegat
C. Baltic Sea
D. Bay of Biscay

35. In which one of the following ocean voyages would a ship cover the SHORTEST distance? 35.A

A. New York to London
B. New York to Rio de Janeiro
C. Rio de Janeiro to London
D. Valparaiso to New York

36. Which one of the following countries touches upon or lies south of the Equator? 36. D
   A. Philippines
   B. Thailand
   C. India
   D. Indonesia

37. The MAJOR tundra areas of Asia are to be found in the 37. C
   A. Iranian-Afghanistan-Punjab regions
   B. areas of Mongolia and Manchuria
   C. Siberian Provinces
   D. Yangtze and Hsiang River valleys

38. Japan has been able to sustain a population of 95 million people because 38. C
   A. her natural mineral resources permit the export of surplus iron and zinc
   B. sufficient quantities of rice and fish are produced to feed the population
   C. she exports enough manufactured goods to pay for imports of food and raw materials
   D. the population requires less caloric intake than western populations

39. In which of the following is the country INCORRECTLY matched with its capital city? 39. C
   A. Laos - Vientiane
   B. Malaysia - Singapore
   C. Turkey - Ankara
   D. Thailand - Bangkok

40. Because of the deflective influence of the earth's rotation, the trade winds blow in a(n)____ direction ______ the Equator. 40. C
   A. westerly; away from
   B. westerly; toward
   C. easterly; toward
   D. easterly; away from

41. Which one of the following is TRUE of Russian rivers? 41. B
   A. Water transport carries as much freight as do Soviet railroads.
   B. Most of the major rivers do not reach the open seas except by roundabout canal connections.
   C. The most heavily used rivers are in Siberia.
   D. The northward flow of the major rivers in European Russia handicaps freight shipments.

42. In which one of the following do both cities have a marine climate? 42. C
   A. London - San Diego
   B. Lisbon - Atlanta
   C. Paris - Seattle
   D. Berlin - Salt Lake City

43. On March 21 and September 22, the sun's rays are vertical at the 43. B
   A. Arctic Circle
   B. Equator
   C. Tropic of Capricorn
   D. Tropic of Cancer

44. The rainfall and physiography of the Ukraine are MOST similar to those of 44. A
A. western Nebraska
B. New England
C. Louisiana
D. the Pacific Northwest

45. Large-scale cotton growing in the United States is characteristic of 45. A
A. western Texas and southern Oklahoma
B. the Tennessee Valley
C. the Mississippi-Arkansas-Red River Valleys
D. the Southern Piedmont Region

46. Which one of the following is TRUE of the geography of South America? 46. D
A. The cultivable land area per person is lower than in Europe or Asia.
B. The physical contours of the continent enhance the possibilities of economic unity and cooperation.
C. Its total area is greater than that of North America.
D. The largest percentage of its total land area is in plains and grasslands.

47. Which one of the following is TRUE of Africa south of the Sahara? 47. B
A. There is a wide coastal plain with important production lands.
B. The abrupt rise of a level plateau makes access to the interior difficult.
C. Most of the major rivers which lead into the interior are navigable.
D. There are many natural harbors on the west African coast.

48. The floor of the deep ocean 48. A
A. has a topography characterized by a great variety of irregularities
B. is a featureless plain
C. is traversed by submarine canyons extending from the margins of adjacent continents
D. is underlain by vast masses of granitic material

49. Neap tides, at any given locality, represent the month's _____ high tide and _____ low tide. 49. C
A. highest; lowest
B. highest; highest
C. lowest; highest
D. lowest; lowest

50. Of the following, an ocean current that flows southward is the 50. D
A. Gulf Stream
B. Benguela Current
C. Peru Current
D. California Current

KEY (CORRECT ANSWERS)

1. B	11. D	21. B	31. C	41. B
2. A	12. A	22. D	32. A	42. C
3. C	13. D	23. B	33. B	43. B
4. D	14. C	24. D	34. D	44. A
5. A	15. D	25. D	35. A	45. A
6. A	16. A	26. B	36. D	46. D
7. D	17. A	27. A	37. C	47. B
8. A	18. A	28. D	38. C	48. A
9. B	19. C	29. A	39. C	49. C
10. B	20. A	30. D	40. C	50. D

TEST 3

DIRECTIONS: Each question or incomplete statement is followed by several suggested answers or completions. Select the one that BEST answers the question or completes the statement. *PRINT THE LETTER OF THE CORRECT ANSWER IN THE SPACE AT THE RIGHT.*

1. Which of the following statements about the Po Valley is TRUE? 1. A
It
   A. is highly developed both industrially and agriculturally, constituting the most economically advanced part of Italy
   B. is rich in soil, rainfall, and water power, but undeveloped industrially because it lacks coal and iron
   C. encompasses the most highly developed industrial complex in Italy because it alone produces an adequate supply of coal and iron
   D. is industrially advanced but lags in agricultural techniques

2. Which statement is NOT true of the river system of Russia? 2. C
   A. It is believed to be the most extensive in the world
   B. It runs crosscurrents to the channels of trade
   C. It carries more tonnage than Russian railroads
   D. The Volga River supplies cheap transportation of products to the Russian markets.

3. Which one of the following factors accounts for the expansion of Brazilian manufacturing? 3. A
   A. Vast iron ore deposits
   B. Economic penetration of the Amazon
   C. High grade coal and oil deposits
   D. A transportation network of rivers and railroads that extends to most parts of the country

4. Which one of the following tropical rainforest areas has a population distribution that is DIFFERENT from the others? 4. D
   A. Borneo
   B. Congo Basin
   C. Amazon Basin
   D. Java

5. Which one of the following American cities has climatic conditions comparable to the foreign city with which it is paired? 5. A
   A. Los Angeles - Nice
   B. Miami - Tel Aviv
   C. New Orleans - Athens
   D. New York - London

6. What proportion of Russia lies above the latitude on which New York City is located? Approximately 6. D
   A. 45% B. 60% C. 75% D. 90%

7. Which one of the following cities is NOT on the Fall Line? 7. A
   A. Atlanta, Georgia B. Richmond, Virginia
   C. Columbus, Georgia D. Columbia, South Carolina

8. Which one of the following was the MAIN factor in the establishment of Canada's aluminum industry? 8. C
   A. Proximity to markets
   B. Availability of domestic bauxite
   C. Availability of low cost power
   D. Large domestic demand for Canadian manufactures

9. Which one of the following CORRECTLY pairs the author and the title of a book on conservation? 9. C
   A. Rachel Carson - WASTEMAKERS
   B. José de Castro - ROAD TO SURVIVAL
   C. Fairfield Osborn - OUR PLUNDERED PLANET
   D. William Vogt - SEA AROUND US

10. Which one of the following scholars emphasized the influence of economic rather than geographic factors on society? 10. A
    A. Charles A. Beard B. Preston James
    C. Ellen Churchill Semple D. Frederick Jackson Turner

11. Which one of the following factors induced more Europeans to settle in British East Africa than in British West Africa? 11. D
    A. Availability of mineral resources
    B. Closeness of world trade centers
    C. Friendliness of natives
    D. Topography and climate

12. Hawaii and Puerto Rico are MOST dissimilar with respect to 12. C
    A. average temperature range
    B. chief crop
    C. extent of manufacturing
    D. latitude

13. If you knew the longitude of two islands located at 0° latitude, you could figure out all of the following EXCEPT 13. D
    A. distance between them in miles
    B. shortest route between them
    C. time difference between them
    D. their climate

14. Which one of the following is the MOST important region for fisheries? 14. C
    A. Indian Ocean
    B. Mediterranean
    C. North Pacific
    D. South Atlantic coastal region

15. Which one of the following crops depletes the soil LEAST? 15. C
A. Corn B. Cotton C. Hay D. Tobacco

16. Which one of the following climatic regions has the GREATEST seasonal variation in temperature? 16. D
A. Desert B. Humid continental
C. Mediterranean D. Taiga

17. Which one of the following can be grown in almost any climate? 17. D
A. Corn B. Cotton
C. Fruits and vegetables D. Wheat

18. Which one of the following cities BEST fits the designation of *the Minneapolis of Europe*? 18. B
A. Berlin B. Odessa C. Vienna D. Warsaw

19. Which one of the following is MOST accurately represented on a Mercator projection? 19. A
A. Equatorial areas B. Meridians of longitude
C. Polar areas D. Shapes of continents

20. Approximately what percentage of the earth's surface is covered by bodies of water? 20. C
A. 30% B. 50% C. 75% D. 90%

Questions 21-23.

DIRECTIONS: Questions 21 through 23 are to be answered on the basis of the map below.

The area indicated on the map below is part of a continent.

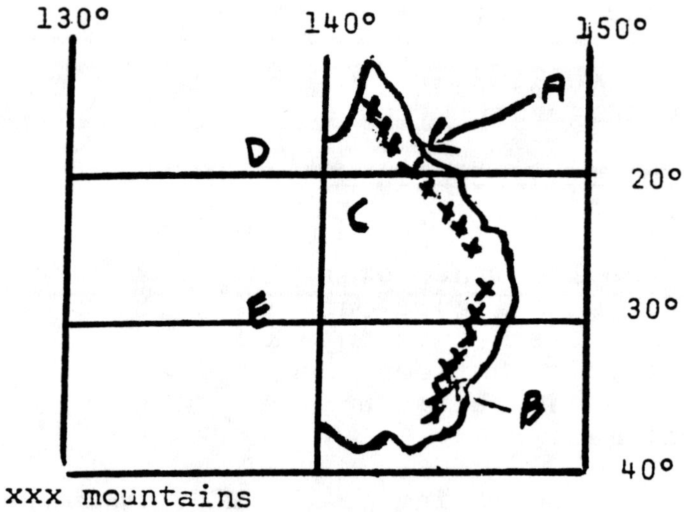

21. To which country is the area indicated on the map CLOSEST? 21. D
A. Japan B. Labrador
C. Madagascar D. New Zealand

22. Which area is CORRECTLY associated with its agricultural product? 22. A
   A. A - pineapples B. B - sugar
   C. C - potatoes D. D - orchards

23. Which area is CORRECTLY associated with its land use? 23. A
   A. B - mixed farming B. C - dairy land
   C. D - lumbering D. E - grazing

24. To which one of the following cities would the accompanying chart apply? 24. C
   A. Cairo, Egypt
   B. Capetown, Union of South Africa
   C. Liverpool, England
   D. New York City, N.Y.

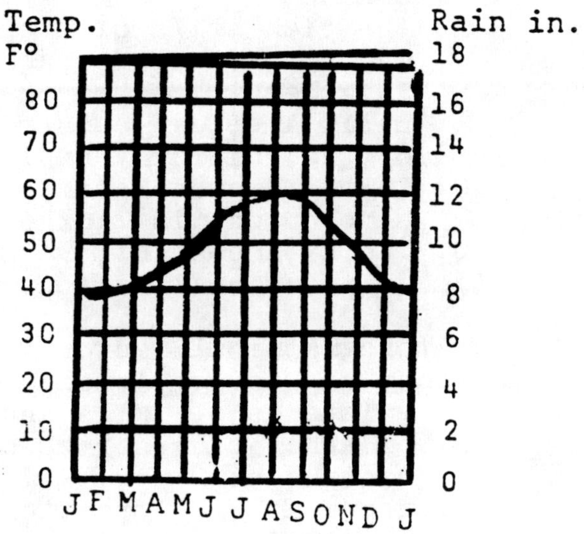

25. Which one of the following CORRECTLY pairs a natural resource with an area in which it is found in large quantities? 25. C
   A. Copper - Mesabi Region, United States
   B. Iron - Atacama Desert, Chile
   C. Nickel - Sudbury, Canada
   D. Oil - Minas Gerais, Brazil

26. Which one of the following is TRUE with regard to the Sahara Desert? 26. A
This region
   A. has large oil deposits
   B. has a minimum temperature of 85°F
   C. is devoid of plant and animal life
   D. is an unbroken plain of sand

27. Which one of the following is NOT a characteristic of the Middle East? 27. A
   A. Climatic conditions are uniform.
   B. The regions with water are densely populated.
   C. 80-85% of the people are engaged in agriculture.
   D. 75% of the world's known oil resources are located here.

28. Which one of the following is TRUE of Africa south of the Sahara? 28. A
   A. Ninety percent of the area has a tropical climate.
   B. Its major occupation is mining.
   C. Two-thirds of the area remains under European colonial rule.
   D. It approximates the size of the United States.

29. Which one of the following statements about Latin America is INCORRECT? 29. A
   A. The rich soils of tropical Latin America offer a major resource for agricultural expansion.
   B. In general, the major commercial products of the Latin American countries use only a small percentage of the gainfully employed workers.
   C. In the Amazon Basin, a major handicap to economic development is lack of manpower.
   D. The high cost of transportation is a serious deterrent to industrialization.

30. The topographical factor that has MOST impeded exploration of the interior of Africa is the 30. A
   A. existence of falls and cataracts on the large rivers
   B. presence of a series of immense rift valleys on the east coast
   C. tropical rainforest of the Congo region
   D. existence of immense stretches of arid and desert lands in Northern Africa

31. The Middle East, with its wealth of oil deposits, 31. A
   A. is hampered by the scarcity of cultivable land in the area
   B. is self-sufficient with respect to coal and iron
   C. is considered potentially capable of industrial development comparable to that of Western Europe
   D. has surpassed the United States as the world's leading oil production region

32. The summer monsoon is USUALLY accompanied by heavy rainfall because 32. D
   A. coastal mountain ranges prevent the moist air from rising
   B. the land cools much more rapidly than the surrounding waters
   C. monsoon winds generally are part of the prevailing Westerlies
   D. light, warm, moist air rises as it passes over hot lands

33. Which one of the following statements MOST accurately describes a major difference between Australia and New Zealand? 33. A
   A. Australia is drought-ridden; New Zealand has considerable rainfall.
   B. Australia is densely populated; New Zealand is sparsely populated.

C. The bulk of Australia's population comes from Western Europe; the bulk of New Zealand's population comes from the Asiatic mainland.
D. Australia's industries are primarily extractive; New Zealand's economy is geared to grazing.

34. In which one of the following groups are the cities CLOSEST to the same parallel of latitude? 34. C
A. New York - Berlin - London - Warsaw
B. New York - Berlin - Frankfort - Paris
C. New York - Istanbul - Madrid - Naples
D. Antwerp - London - Vienna - Warsaw

35. Which one of the following is NOT used by cartographers to represent topographical relief? 35. A
A. Isobars B. Contour C. Hachures D. Color

36. Within an hour, a given point on the earth passes through 36. A
A. 15 degrees
B. 30 degrees
C. a varying number of degrees in accordance with changes in latitude
D. 30 degrees at the Tropic of Cancer

37. Through which body of water would an oil tanker pass FIRST en route from Kuwait to France via the Suez Canal? 37. C
A. Arabian Sea B. Indian Ocean
C. Persian Gulf D. Red Sea

38. Which one of the following would be the LAST port of call of a vessel using the St. Lawrence Seaway from the Gulf of St. Lawrence to the iron ore fields at the western tip of Lake Superior? 38. B
A. Cleveland B. Duluth
C. Milwaukee D. Toronto

39. For MOST of their length, the Tigris and Euphrates Rivers flow through 39. B
A. Iran B. Iraq
C. Saudi Arabia D. Syria

40. Which one of the following ports is normally ice-blocked during the winter? 40. B
A. Hammerfest, Norway B. St. Petersburg, Russia
C. Murmansk, Russia D. Vladivostok, Russia

41. Which one of the following has BOTH a larger area and a larger population than the other three? 41. B
A. Argentina B. Brazil C. Chile D. Mexico

42. Which one of the following items describes the MAJOR use of land in the United States today? 42. A
A. Pasture and grazing land
B. Forests and woodland, not pastured
C. Cropland
D. Urban areas, parks, highways, and other public facilities

43. Which one of the following statements about the geography of Canada is INCORRECT? 43. B
   A. Most of the country lies north of 50° latitude.
   B. Canada has a population roughly twice that of New York State.
   C. Most of Canada's topography is hostile to permanent settlement.
   D. Since World War II, there have been important discoveries of oil, iron ore, and non-ferrous metals.

44. Which one of the following factors necessary for industrialization does Japan lack? 44. D
   A. Manpower
   B. Power resources
   C. Suitable climate
   D. Iron ore deposits

45. Which one of the following types of climate is characterized by cool, damp summers and mild, wet winters? 45. C
   A. Tropical rainforest
   B. Mediterranean
   C. Marine west coast
   D. Humid continental

46. There is a significant difference between the North and South Polar regions with respect to 46. C
   A. temperature
   B. vegetation
   C. altitude
   D. precipitation

47. The OUTSTANDING advantage of the Mercator projection is utilized in 47. A
   A. charting nautical courses
   B. showing the distribution of economic resources
   C. giving accurate representation of the overall shape of continents
   D. indicating the earth's major topographical features

48. The OUTSTANDING topographic feature of Eurasia is 48. A
   A. a great east-west barrier of mountains, plateaus, basins, and deserts
   B. an immense north-south plain flanked by highlands
   C. an absence of navigable rivers and choice harbors
   D. its location in predominantly tropical and equatorial latitudes

49. Of the following, the country that is situated ENTIRELY within the North Temperate Zone is 49. C
   A. Norway
   B. Sweden
   C. Iceland
   D. Finland

50. Of the following statements about the area of the former Soviet Union, the one which is TRUE is: It was _____ than the combined area of _____. 50. D
   A. less; Brazil and India
   B. less; U.S.A. and Canada
   C. less; China and the U.S.A.
   D. greater; China and Canada

KEY (CORRECT ANSWERS)

1. A	11. D	21. D	31. A	41. B
2. C	12. C	22. A	32. D	42. A
3. A	13. D	23. A	33. A	43. B
4. D	14. C	24. C	34. C	44. D
5. A	15. C	25. C	35. A	45. C
6. D	16. D	26. A	36. A	46. C
7. A	17. D	27. A	37. C	47. A
8. C	18. B	28. A	38. B	48. A
9. C	19. A	29. A	39. B	49. C
10. A	20. C	30. A	40. B	50. D

TEST 4

DIRECTIONS: Each question or incomplete statement is followed by several suggested answers or completions. Select the one that BEST answers the question or completes the statement. *PRINT THE LETTER OF THE CORRECT ANSWER IN THE SPACE AT THE RIGHT.*

1. The Campos of Brazil differ MOST markedly from the Llanos of Venezuela in 1. A
   A. topography B. climate
   C. natural vegetation D. land utilization

2. Which one of the following generalizations concerning the pattern of agriculture in South America is MOST accurate? 2. C
   A. Subsistence agriculture is characteristic of the lowlands near the coast, while the commercial plantations are in the tropical highlands.
   B. The vast expanse of the Andean highlands, generally not usable for agriculture, crowds plantations and subsistence farms into the coastal lowlands.
   C. The principal plantation areas are along the north and eastern coasts, while the important subsistence agriculture regions are in highland areas.
   D. With the exception of the coffee plantations in the highlands, most plantation agriculture and practically all subsistence agriculture are carried on in the lowland regions.

3. The agricultural products of the Central Valley of Chile MOST closely resemble those of 3. C
   A. southern Iran
   B. Vancouver Island
   C. the Cape district of Africa
   D. the coastal plain of Queensland

4. Which one of the following can be MOST correctly cited as a geographic reason for the delayed development of Africa? The 4. A
   A. nature of the topography
   B. dense forests that cover the major part of the continent
   C. generally unfavorable climatic conditions throughout the continent
   D. general absence of fertile soil in regions otherwise suitable for agriculture

5. Which one of the following deserts owes its aridity to a combination of causal factors different from that of the other three? 5. D
   _____ Desert.
   A. Sahara B. Atacama
   C. Australian D. Gobi

6. Which of the following regions of the world can PROPERLY be grouped as illustrating basically similar utilization of the natural environment? 6. A

I. Northern Africa
II. Patagonia
III. Sinkiang
IV. Arabia
V. Mongolia

The CORRECT answer is:

A. I, III, IV, V
B. II, IV, V
C. III, V
D. I, III, IV

7. Which group consists entirely of European countries wherein the percentage of population engaged in agriculture is less than the percentage of the population of the United States that is engaged in agriculture? 7. C

A. France, Norway, Germany
B. England and Wales, Italy, Czech Republic
C. Belgium, Holland, Switzerland
D. Portugal, Sweden, Austria

8. Starting at the Canadian border and traveling south along the Mississippi Valley, one would pass the following agricultural belts: 8. B

I. Corn
II. Diversified and general farming
III. Dairy

The CORRECT order in which these belts occur is:

A. I, II, III
B. III, I, II
C. II, III, I
D. III, II, I

9. Which one of the following continents is CORRECTLY matched with a desert located in it? 9. D

A. Asia - The Kalahari Desert
B. North America - The Patagonia Desert
C. Africa - The Gobi Desert
D. Australia - The Great Sandy Desert

10. Which one of the following statements BEST represents the geographic character of the Middle East? 10. D

A. Contrary to popular impression, it is not strategically located for the merging of continents and oceans.
B. Although oil is the best known resource, other mineral resources are of greater importance.
C. The greatest limiting factor to cultivation is not water but fertile oil.
D. While the image of the area is one of unrelieved aridity and deserts, there are significant variations of climate and topography.

11. Of the following groups of products, the one MOST likely to be found in Southeast Asia is rice, 11. A

A. teak, rubber, petroleum
B. wheat, copra, palm oil
C. wheat, copra, mango
D. corn, tobacco, kapok

12. The teacher places the following terms on the chalkboard: plateau, mountainous, source of a great river, Humboldt Current, volcanoes.
The region of the world she is presenting to the class is PROBABLY the 12. B
   A. east coast of South America
   B. west coast of South America
   C. west coast of North America
   D. subcontinent of India

13. An ACCURATE conclusion which can be drawn by a class studying a map of South America is that 13. A
   A. the Andes present a significant obstacle to the development of an effective network of roads and railroads
   B. river systems tend to serve as barriers to communication
   C. mountains prevent the development of large urban centers
   D. tropical rainforests cover most of western South America

14. An ADVANTAGE to students and teachers in using a Mercator projection is that it shows 14. B
   A. all land areas without distortions
   B. all directions correctly and accurately
   C. only distortions near the Equator
   D. Great Circle routes most accurately

15. Weather and climate are closely interrelated, but DIFFER in that 15. B
   A. weather is the result of the behavior of the sun, while climate is chiefly influenced by land and water distribution
   B. climate is the average weather over a long period of time
   C. weather has a greater effect upon man's economic activities than does climate
   D. man can control the vicissitudes of the climate but has no influence over weather

16. The four seasons of the year occur because of the 16. A
   A. tilted axis of the earth, which maintains a constant orientation in space as the earth revolves about the sun
   B. change in distance from the sun which the earth experiences in its revolution around the sun
   C. occurrence of the fall and spring equinoxes and the winter and summer solstices
   D. rotation of the earth which results in periods of daylight and night

17. Of the following statements about sowing and harvesting wheat, the CORRECT one is that 17. A
   A. winter wheat is sown in the fall, thrives in the cool moist winter, ripens and is harvested in the late, dry summer
   B. spring wheat is sown in the spring and harvested in the early summer, thereby allowing for the planting of summer wheat
   C. winter wheat is sown during the winter, grows during the spring and summer, and is harvested in late fall, prior to the planting of the next winter crop
   D. spring wheat is sown in early spring and harvested in late spring, thereby allowing for the planting of a different crop in the summer

18. The problem of providing food for the world's population in the future 18. D
   A. will not be important because the birth rate declines in all nations as they develop, industrialize, and modernize
   B. is unsolvable unless drastic action is taken through the science of eugenics to reduce the world's population
   C. will be solved by the inevitable havoc of a nuclear war
   D. may be met by such measures as extracting protein from non-edible plants, increased use of protein-rich fish meal, and the domestication of wild animals

19. Of the following nations, the one which shares a border with both China and Tajikistan is 19. C
   A. Iran
   B. India
   C. Afghanistan
   D. Pakistan

20. The lack of modern economic development of the Amazon Basin of Brazil is due to all of the following reasons EXCEPT 20. D
   A. lack of development capital
   B. lack of a developed infrastructure
   C. failure to find mineral deposits
   D. failure of government policy in recognizing the area's wealth

Questions 21-25.

DIRECTIONS: Questions 21 through 25 are to be answered on the basis of the map below. The area indicated on the map is part of a continent.

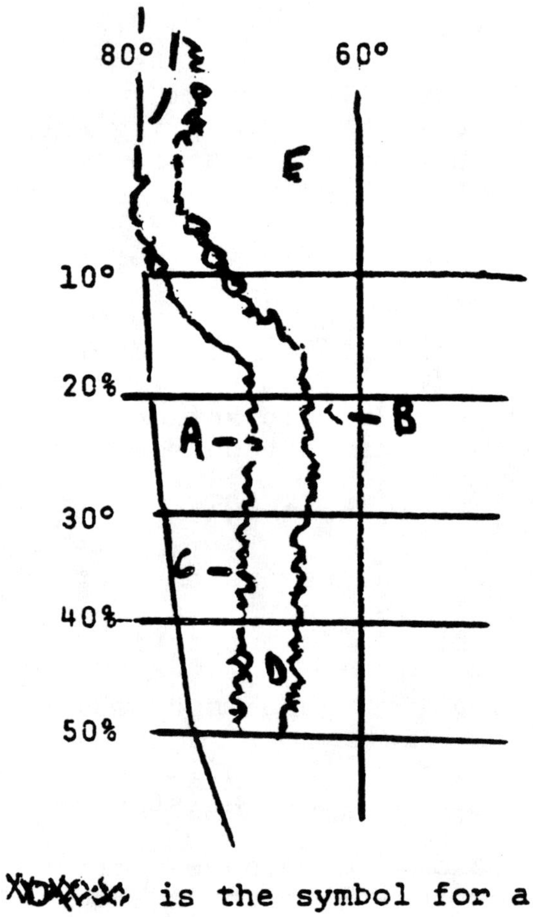

is the symbol for a mountain range

21. To which one of the following countries is the area indicated on the map CLOSEST? 21. C
    A. Indonesia B. Nigeria C. Panama D. Greenland

22. Which letter is CORRECTLY matched with the climate? 22. C
    A. A - tropical rainforest B. C - desert
    C. D - marine D. E - humid continental

23. Where would the THICKEST forests be found? 23. A
    A. E B. B C. C D. A

24. Which area is CORRECTLY matched with its product? 24. D
    A. A - forests B. C - rice
    C. E - cotton D. D - sheep

25. Which one of the following factors is MOST favorable to the economic development of the region on the map? 25. B
A. Proximity to markets
B. Export trade
C. Natural as well as man-made transportation facilities
D. Diversified economy

26. Which one of the following has been the MAJOR handicap in the production of iron ore from the Mesabi Range? 26. D
A. Difficulty and expense of mining the iron ore
B. Inadequacy of port and transportation facilities
C. Depletion of usable ores
D. Absence of ore mining in the winter

27. Russia and the Ukraine are the leading world producers of which of the following commodities? 27. A
A. Rye, hemp, flax, fiber B. Rice, corn, jute
C. Henequen, silk, barley D. Tobacco, tea, soybean

28. Africa is a leading producer of which of the following commodities? 28. B
A. Rice, hemp, jute
B. Sisal, cacoa, palm oil
C. Rubber, tobacco, coffee
D. Corn, flax fiber, yerba mate

29. Which one of the following synthetic products has been MOST competitive in the economy of Southeast Asia, particularly Vietnam and Malaysia? 29. B
A. Nylon rope B. Buna rubber
C. Saccharin D. Polyethylene

30. Which one of the following has hampered the industrial development of the Congo Republic? 30. A
A. Expensive transportation
B. Absence of water power or coal
C. Uniformity of tropical rainforest
D. Absence of agriculture

31. Which one of the following was the LEAST important deterrent in the penetration of the interior of sub-Sahara Africa? 31. C
A. Plateau topography B. Regular coastline
C. Climate D. Navigability of rivers

32. Which one of the following publications deals MAINLY with geography? 32. C
A. SCIENTIFIC AMERICAN B. SOCIAL EDUCATION
C. FOCUS D. REPORTER

33. Which one of the following factors is UNFAVORABLE to the development of manufacturing in the Rocky Mountains area? 33. D
A. Absence of coal
B. Lack of government assistance
C. Lack of raw materials
D. Distance from markets

34. Which one of the following is NOT correctly matched with an important resource for which it is known? 34. D
   A. Karaganda - coal
   B. Ukraine - iron
   C. Caspian Sea area - oil
   D. Siberian Tundra - aluminum

35. Which one of the following statements is TRUE regarding the geography of India? 35. B
   A. The average annual temperatures increase from south to north, with local elevations offsetting the effect of latitude.
   B. The Ganges and coastal plains are more suitable for agriculture than the South Deccan Plateau.
   C. India supplies jute to Pakistan's factories.
   D. The monsoons have no economic value because the rain is too abundant.

36. In which period of the year would the peasants in Vietnam face the GREATEST problem from rain monsoons? 36. C
   A. January B. March C. July D. November

37. Which one of the following factors is MOST favorable to the industrial development of China? 37. C
   A. Extensive forest reserves
   B. Accessibility of water power
   C. Presence of mineral resources
   D. Extensive railroad network linked to its river and canal system

38. Which area in Africa is CORRECTLY matched with the physical feature associated with it? 38. A
   A. Union of South Africa - Great Escarpment
   B. Ethiopia - Atlas Mountains
   C. Angola - Great Rift Valley
   D. Central Africa - Sea level altitude

39. Which one of the following climates would be found either north or south of the tropical rainforest in Africa but not in both parts? 39. D
   A. Mediterranean B. Steppe
   C. Savanna D. Marine

40. When it is 10 P.M. in Philadelphia, Pa., what time is it in Greenwich, England? 40. C
   A. 5 P.M. same day B. 1 A.M. next day
   C. 3 A.M. next day D. 11 A.M. same day

41. Industrial development of the Kusnetsk-Sverdlovsk-Karaganda complex was hampered by 41. A
   A. long distances from natural arteries of transportation
   B. lack of coal and iron
   C. unfavorable climate
   D. inaccessibility of raw materials

42. Which one of the following countries has the SMALLEST known reserve of iron ore? 42. D
    A. Brazil B. India
    C. Sweden D. West Germany

43. Which one of the following is NOT a state in the Danube region? 43. C
    A. Slovakia B. Germany
    C. Switzerland D. Serbia

44. The Mekong River runs along the borders of 44. D
    A. Cambodia and Vietnam
    B. Laos and Vietnam
    C. Thailand and Cambodia
    D. Thailand and Laos

45. In which of the following pairs are both countries land-locked? 45. B
    A. Bolivia and Ecuador B. Bolivia and Paraguay
    C. Paraguay and Uruguay D. Uruguay and Peru

46. Which one of the following areas lies in APPROXIMATELY the same range of latitudes as the United States? 46. A
    A. China
    B. India
    C. Scandinavia
    D. Russia

47. Which one of the following does NOT serve as an outlet for rivers flowing through German territory? 47. B
    A. Baltic Sea B. Bay of Biscay
    C. Black Sea D. North Sea

48. Which one of the following geographic factors BEST explains the economic importance of Northern Italy? 48. B
    A. Extensive iron and coal deposits
    B. Favorable agricultural conditions and hydroelectric power
    C. Forests and fisheries
    D. Petroleum and natural gas

49. Which one of the following agricultural practices has a DIFFERENT effect on the land from the other three? 49. D
    A. Contour plowing
    B. Diversified or mixed farming
    C. Strip cropping
    D. Use of clean-tilled crops

50. A study of the world's supply of coal would reveal that 50. A
    I. there is a dearth of coal in the Southern Hemisphere
    II. North America ranks first and Asia second in known reserves of coal
    III. present known reserves of coal indicate the need for developing other sources of power within the next hundred years
    IV. anthracite coal is fairly evenly distributed throughout the Northern Hemisphere

The CORRECT answer is:
A. I, II B. II, III C. I, IV D. III, IV

KEY (CORRECT ANSWERS)

1. A	11. A	21. C	31. C	41. A
2. C	12. B	22. C	32. C	42. D
3. C	13. A	23. A	33. D	43. C
4. A	14. B	24. D	34. D	44. D
5. D	15. B	25. B	35. B	45. B
6. A	16. A	26. D	36. C	46. A
7. C	17. A	27. A	37. C	47. B
8. B	18. D	28. B	38. A	48. B
9. D	19. C	29. B	39. D	49. D
10. D	20. D	30. A	40. C	50. A

TEST 5

DIRECTIONS: Each question or incomplete statement is followed by several suggested answers or completions. Select the one that BEST answers the question or completes the statement. *PRINT THE LETTER OF THE CORRECT ANSWER IN THE SPACE AT THE RIGHT.*

1. Which one of the following has been characteristic of European economy during the past century? 1. B
   A. Its once moderately fertile soils have become exhausted.
   B. Soils unsuited for crops or pasture have been planted to forests.
   C. The loss through depletion of its mineral resources has been more than recouped by the discovery of new mines.
   D. Its sparsely distributed water power has been left largely unharnessed.

2. A person touring the industrial cities of Great Britain would find the wool industry centered at 2. A
   A. Leeds and Bradford B. Liverpool and Manchester
   C. Sheffield and Edinburgh D. Birmingham and Glasgow

3. Which one of the following statements concerning mineral resources in China is NOT true? 3. C
   A. Large deposits of coal are known to exist in China.
   B. China is relatively deficient in iron ore.
   C. Lack of tungsten and antimony resources is an important barrier to industrialization.
   D. Iron ore deposits are generally distant from coking coal deposits.

4. The region in Asia which supports the GREATEST density of agricultural population is 4. D
   A. Malaya B. Ceylon C. China D. Java

5. Which of the following statements concerning Iran are CORRECT? 5. C
   It
   I. is a region of great aridity, much of it actual desert
   II. has common frontiers on the west with Turkey, Syria, Iraq, and Saudi Arabia
   III. has common frontiers with Turkmenistan , Afghanistan, and Pakistan
   IV. borders on the Caspian Sea and the Persian Gulf

   The CORRECT answer is:
   A. I, II, III B. II, III, IV
   C. I, III, IV D. I, II, IV

6. Which one of the following is the geographic factor MOST directly pertinent to the dispute over Kashmir? 6. C
   A. Need for a land bridge between Pakistan and Bangladesh
   B. Absence of coal and mineral resources in India
   C. Pakistan's need for water for irrigation
   D. India's need for raw materials for her jute mills

7. Which of the following may CORRECTLY be considered part of the monsoon lands of Asia? 7. D
   I. Indonesia
   II. Japan
   III. Ceylon
   IV. Northeastern China
   V. Southeastern China

   The CORRECT answer is:
   A. I, III, V
   B. I, II, V
   C. II, IV
   D. All of the above

8. The Pan American Highway System is completed from 8. C
   A. Panama to Santiago
   B. the United States border to Panama
   C. Santiago to Buenos Aires
   D. the United States border to Quito

9. The development of manufacturing in the Piedmont region of the southern part of the United States is due PRIMARILY to the 9. B
   A. presence there of cotton and tobacco for processing
   B. abundant water power and cheap labor of the region
   C. fact that the region is unsuited to the typical southern crops, cotton and tobacco
   D. accessibility of coal for power

10. The Welland Canal connects Lake _____ and Lake _____. 10. A
    A. Erie; Ontario
    B. Superior; Michigan
    C. Huron; Erie
    D. Michigan; Huron

11. Which of the following statements concerning Alaska are TRUE? 11. D
    I. The rugged topography of the Alaskan panhandle district has made it a sparsely populated region despite the fact that its climate is milder than the other regions of Alaska.
    II. The fishing industry employs more labor than the mining industry, but the products of mining exceed those of fishing in value.
    III. The food exports from the United States to Alaska are only those crops and products that cannot be produced in Alaska because of handicaps of soil and climate.
    IV. The Matanuska Valley experiment proved the practicability of dairying and vegetable farming in Alaska.
    V. Reserves of petroleum, anthracite coal, and bituminous coal are known to exist in Alaska.

The CORRECT answer is:
I, II, III, V B. I, II, IV, V
C. I, III, IV D. II, III, IV, V

12. Recent investigations in Canada have revealed significant deposits of 12. C
A. nickel B. copper C. iron ore D. zinc

13. Which of the following factors account for the comparative isolation of the lower St. Lawrence Valley? 13. C
I. The river is ice-bound for five to six months a year.
II. Subsistence farming has increased in the last generation.
III. There are few good harbors along the river.
IV. Heavy snows make transportation difficult in winter.
V. Deforestation has caused a rapid decline of the pulp and paper industry.

The CORRECT answer is:
A. I, II, III, V B. I, II, IV, V
C. I, III, IV D. II, III, IV, V

14. Which one of the following statements is CORRECT concerning Arctic America? 14. B
A. Some plant life grows but the environment permits production of nothing of economic value.
B. The mid-summer days become quite warm or even hot.
C. The lands are cold deserts.
D. The reindeer is the only animal that exists in large numbers in this region.

15. Of the following, the MOST complete and accurate description of the pattern of land utilization on the Great Plains of the United States is 15. C
A. wheat farming in the region east of the one hundredth meridian, commercial grazing west of that line, wheat, forage crops and sugar beets in the irrigable districts
B. commercial grazing throughout the region except in years when unusual demand for wheat makes production of that crop profitable using dry farming methods
C. commercial grazing in the sections not suited for crops, wheat farming in the humid margins, forage crops and sugar beets in the irrigable districts
D. commercial grazing concentrated in the southern part of the region, wheat farming concentrated in the more humid northern part, forage crops and sugar beets in the irrigable districts

16. The Corn Belt of the United States 16. B
I. is characterized by farms that average about 400 acres in size
II. is bounded on the north by the hay and dairying belt
III. produces hay, wheat, oats, soybeans, and tobacco, as well as corn

IV. is, in part, the prairie region reached by early settlers from the east
V. is a region that employs crop rotation to maintain soil fertility

The CORRECT answer is:
A. I, II, III
B. II, III, IV, V
C. III, IV, V
D. I, II, V

17. The spring wheat region of the United States includes parts of the states of 17. A
A. Minnesota, North Dakota, South Dakota, Montana
B. Texas, Oklahoma, Kansas, Nebraska
C. West Virginia, Ohio, Indiana, Illinois
D. Washington, Oregon, Idaho, Montana

18. Which of the following CORRECTLY associate physical features of the Intermontane area of the United States with related irrigated oases? 18. D
I. Colorado River - Imperial Valley
II. Wasatch Mountains - Eastern Utah
III. Snake River - Southern Idaho
IV. Salt River - Central Arizona
V. San Joaquin River - Great California Valley

The CORRECT answer is:
A. I, III, IV, V
B. I, II, IV
C. II, III, V
D. I, II, III, IV

19. The *Big Inch* pipeline 19. C
A. was abandoned after the close of World War II
B. carries petroleum from Texas and Oklahoma to New York and Philadelphia
C. carries natural gas from Texas to New York and Philadelphia
D. carries petroleum from Ohio and Indiana to Chicago

20. The steel industry is generally located near coal PRINCIPALLY because 20. C
A. most coal mines are in the vicinity of seaports
B. the iron needed in the manufacture of steel is usually found in deposits close to the coal mines
C. the bulk of the coal needed to make steel is many times the bulk of iron
D. labor is cheaper in coal areas than in iron areas

21. The technique MOST frequently employed in dry farming is 21. B
A. silting
B. fallowing
C. flood-watering
D. irrigating

22. The diurnal temperature range is GREATEST in 22. A
A. tropical deserts
B. humid continental regions
C. areas of Mediterranean climate
D. tropical rainforests

23. Possession by an area of resources of salt, sulphur, lime, and coal makes possible industrial development in 23. B
A. plastics
B. chemicals
C. glassware
D. metal alloys

24. In connection with which of the following products has the rivalry between producers utilizing basically different climate environments been MOST prolonged and significant? 24. D
A. Coffee
B. Beef cattle
C. Natural rubber
D. Sugar

25. Transatlantic planes flying from North America to Europe follow the _____ and returning to America follow the _____. 25. A
A. Westerlies; Northeast Polar Winds
B. Westerlies; Southeast Trade Winds
C. Northeast Polar Winds; Westerlies
D. Westerlies; Northwest Polar Winds

26. Which of the following are characteristics of regions of Mediterranean climate? 26. D
I. They are located in the middle latitude areas.
II. They have drought in winter and rain in summer.
III. Wheat is a typical crop.
IV. They employ summer irrigation where water is available.
V. They have an absence of trees.

The CORRECT answer is:
A. III, IV, V
B. I, II, V
C. I, II, III
D. I, III, IV

27. The fact that New York early surpassed Boston, Philadelphia, and Charleston in importance as a port is BASICALLY due to 27. B
A. the superiority of New York harbor over other east coast ports
B. its early ability to tap a larger hinterland than other east coast ports
C. the concentration of shipping interests that developed from the fishing industry
D. the network of railroads that developed around the city

28. Which one of the following would a land traveler notice in moving from a rainforest to a savanna climate? A change from 28. A
A. lower trees and thick underbrush to taller trees with little undergrowth
B. an essentially one-season year to alternating wet and dry seasons
C. a lower to a higher latitude
D. lateritic soils that are more leached to soils that are less leached

29. The type of physical map that uses a series of lines to depict such surface configurations as degree of slope, relative relief, and elevation is called a(n) _____ map. 29. B

A. altitude tint
B. contour
C. geodesic survey
D. shaded relief

30. In which one of the following is a map projection paired with its MOST useful purpose? 30. B

A. Mercator - Crop production
B. North polar equidistant - Air travel routes
C. Conic - Colonial empires
D. Goode's homolosine - Navigation routes

31. Which one of the following parallels of latitude crosses extensive deserts? 31. B

A. The Equator
B. 20° north latitude
C. 60° north latitude
D. 60° south latitude

32. The GREATEST beef exporting regions today are in the 32. B

A. middle latitudes of the Northern Hemisphere
B. middle latitudes of the Southern Hemisphere
C. tropics
D. sub-tropical areas

33. The length of a degree of latitude is everywhere APPROXIMATELY _____ miles. 33. C

A. 15
B. 28
C. 69
D. 90

34. Which one of the following is CORRECT regarding the distance between degrees of longitude? It 34. C

A. does not vary at any place on the earth's surface
B. is greater at the poles than at the Equator
C. is greatest at the Equator
D. is greater in the Northern Hemisphere than in the Southern Hemisphere

35. The Island of Formosa is in APPROXIMATELY the same latitude as the Island of 35. A

A. Cuba
B. Australia
C. Hawaii
D. Madagascar

36. Two nations border on North Korea. They are 36. B

A. China and Vietnam
B. China and Russia
C. Russia and Japan
D. China and India

37. In the production of which of the following groups of strategic materials is the United States MOST deficient? 37. B

A. Zinc, lead
B. Chromium, manganese
C. Copper, molybdenum
D. Aluminum, iron ore

38. Which of the following basic materials is NOT correctly paired with the states that are important producers of the commodity? 38. B
A. Crude petroleum - California, Texas
B. Iron ore - Minnesota, Mississippi
C. Steel - Ohio, Pennsylvania
D. Wheat - North Dakota, Iowa

39. Which of the following statements are TRUE about recent shifts in the population of the United States? 39. B
I. The states on the Pacific coast are the fastest growing in the country.
II. Since the 1940 census, the Middle Atlantic States have been losing population.
III. California is now estimated to be the first or the second most populated state in the United States.
IV. In 1968, more than half of the American population was living in urban areas.
V. In the last decade, our birth rate has been rising while our death rate has been falling.

The CORRECT answer is:
A. IV, V
B. I, III, IV, V
C. I, II, IV
D. I, IV, V

40. Which of the following statements about the State of Israel are TRUE? 40. A
I. Its area, as determined by the original UN Partition Plan, is only about half the size of Vermont.
II. Haifa is the terminus of an important oil line running from Iraq.
III. Its bordering countries are Egypt, Lebanon, Syria, Saudi Arabia, and Hashemite Jordan Kingdom.
IV. Oranges are an important export crop of Israel.
V. In climate and topography, it resembles closely the southern part of California.

The CORRECT answer is:
A. I, II, IV, V
B. II, III, V
C. I, II, IV
D. I, IV, V

41. Which three countries contain more than half of the entire known oil reserves of the Middle East? 41. A
I. Lebanon
II. Iraq
III. Saudi Arabia
IV. Iran
V. Israel

The CORRECT answer is:
A. II, III, IV
B. I, III, V
C. I, II, III
D. II, IV, V

42. Which of the following statements about Canada are TRUE? 42. B
I. In manufacturing, Canada is surpassed today only by the United States, Great Britain, and Russia.
II. Canada's population is smaller than that of the United States but larger than that of any other nation in the western hemisphere except Brazil.
III. Canada surpasses continental United States in the following: area, wood exports, nickel, asbestos.
IV. Canada normally sells more to the United States than she buys from the United States.
V. Canada normally sells more to the United Kingdom than she buys from the United Kingdom.

The CORRECT answer is:
A. II, III, IV
B. I, III, V
C. II, III, V
D. I, II, V

43. Which of the following countries secures the HIGHEST yield of wheat per acre of land? 43. A
A. England
B. Argentina
C. Russia
D. United States

44. Which of the following comparisons of Great Britain with Japan are TRUE? 44. B
I. Both nations must today import most of their food supply.
II. Since World War II, both have been faced by the loss of important markets and sources of raw materials.
III. Japan's population density per square mile is from two to three times that of Great Britain.
IV. Both are large importers of raw cotton and manufacturers of cotton textiles.
V. Before World War II, both nations ranked about the same in the production of iron and steel.

The CORRECT answer is:
A. I, III, V
B. I, II, IV
C. II, IV, V
D. II, IV

45. Which of the following statements are TRUE about the territories near the Arctic Circle? 45. B
I. Their growing significance stems more from military reasons than from economic factors.
II. In most of the areas of Greenland, Alaska, and Iceland, the climate is practically the same.
III. Greenland is the largest island in the world.
IV. The shortest air routes between the United States and Europe and between the United States and Asia pass over the Arctic.
V. The largest cities north of the Arctic Circle lie in Russia.

The CORRECT answer is:
A. I, II, IV
B. I, III, IV, V
C. II, III, V
D. I, II, III, IV, V

46. The United States is LARGELY dependent for its atomic materials (uranium and plutonium) on 46. D
   A. India and the Philippines
   B. Peru and Chile
   C. The Union of South Africa and Mexico
   D. Central Africa and Canada

47. Which one of the following statements about South America is NOT correct? 47. C
   A. The coastline is fairly regular and good harbors are generally lacking, except for a few on the east coast.
   B. Bolivia and Paraguay are land-locked.
   C. South America is well-supplied with many mineral resources of which coal is of major significance.
   D. Chile's climate includes rainy west-coast marine in the south, Mediterranean in the center, and desert in the north.

48. To which of the following may the slow development of Brazil be attributed? 48. C
   I. Poor communications
   II. Shortage of capital
   III. Inadequate supplies of coal
   IV. Absence of hydroelectric potential

   The CORRECT answer is:
   A. I *only* B. I, II
   C. I, II, III D. I, II, III, IV

49. Rapids and cataracts near its mouth and an undeveloped hinterland restrict navigation on the _____ River. 49. B
   A. Amazon B. Congo C. Nile D. Zambesi

50. Severe weather in Iraq characterized by very high winds overcast skies and precipitation, normally begin each year during the 50. D
   A. summer B. early winter
   C. fall D. late winter or early spring

KEY (CORRECT ANSWERS)

1. B	11. D	21. B	31. B	41. A
2. A	12. C	22. A	32. B	42. B
3. C	13. C	23. B	33. C	43. A
4. D	14. B	24. D	34. C	44. B
5. C	15. C	25. A	35. A	45. B
6. C	16. B	26. D	36. B	46. D
7. D	17. A	27. B	37. B	47. C
8. C	18. D	28. A	38. B	48. C
9. B	19. C	29. B	39. B	49. B
10. A	20. C	30. B	40. A	50. D

EXAMINATION SECTION

TEST 1

DIRECTIONS: Each question or incomplete statement is followed by several suggested answers or completions. Select the one that BEST answers the question or completes the statement. *PRINT THE LETTER OF THE CORRECT ANSWER IN THE SPACE AT THE RIGHT.*

1. In Indonesia, Thailand, and the Philippines, a large share of the commerce is controlled by — 1. A
   A. Chinese
   B. large landowners
   C. government officials
   D. Moslems

2. Which one of the following accounts for the removal of the rubber-growing industry from Brazil to southeastern Asia and the adjacent islands? — 2. A
   A. More abundant and efficient labor supply is available in the East.
   B. Better climatic and soil conditions exist in the East.
   C. Brazilian plantation systems were already exhausted.
   D. Unstable Asian governments tempted investment capital.

3. Which one of the following statements regarding the *virgin lands* project in Kazakhstan is INCORRECT? — 3. A
   A. It is being conducted in a sub-polar type of climate.
   B. Precipitation in the area is equivalent to that of semi-desert regions.
   C. Huge state farms are being created by the government, manned by young people from all over the Soviet Union.
   D. In the region, transport is limited and population is sparse.

4. Which one of the following statements is NOT true? — 4. D
   A. Europe has larger reserves of iron ore in proportion to its area than has any other continent.
   B. The greater portion of European coal is found in a broad belt following the southern edge of the European plain.
   C. Sweden has important iron ore reserves but limited coal reserves.
   D. Europe's large copper needs are met by its own production.

5. All of the following have presented geographic barriers to the economic development of Africa EXCEPT — 5. D
   A. narrow coastal plains bordering a high plateau
   B. extensive desert and semi-arid areas
   C. poor soil conditions
   D. lack of power resources

6. The Katanga district of Zaire produces cobalt, zinc, and — 6. B
   A. coal
   B. copper
   C. petroleum
   D. rice

7. Which statement is TRUE of the growing of rice, rather than wheat? 7. C
   A. It requires a cool, dry climate.
   B. It is a leading commodity in world trade.
   C. It is generally the product of intensive cultivation.
   D. Its production has been transformed by the use of machinery.

8. All maps have serious shortcomings because 8. C
   A. equal areas cannot be depicted on a map
   B. maps showing geographic land details sacrifice accuracy of shape
   C. distortion and exaggeration of some details must exist
   D. lines of latitude are of the same length

9. Traveling due south from the city of Detroit, Michigan, would bring one to a point in 9. C
   A. Brazil B. Mexico
   C. the Pacific Ocean D. Venezuela

10. A map of the major agricultural regions of the United States shows a(n) 10. B
   A. corn and winter wheat belt around the Great Lakes
   B. fruit and vegetable belt along part of the Pacific coast
   C. hay and dairy region just below the corn belt
   D. irrigated farm belt bounded on the west by the Rocky Mountains

11. Climatic conditions in the southeastern United States MOST closely resemble those of 11. A
   A. southern Brazil B. France
   C. east coast of India D. Morocco

12. Forty percent of the country's labor force is employed in manufacturing in 12. A
   A. Great Britain B. Ireland
   C. Poland D. Union of South Africa

13. Which one of the following is NOT an important source of manganese? 13. A
   A. Canada
   B. India
   C. Union of South Africa
   D. Russia

14. The latitude of Japan is CLOSEST to that of 14. D
   A. Canada B. Russia
   C. the United Kingdom D. the United States

15. A BASIC factor which may seriously retard the rate of Mainland China's industrialization is the fact that 15. C
   A. her large size prevents exploitation of available resources
   B. she is seriously deficient in power resources, notably coal

C. she lacks the requisite amounts of high grade iron ore for heavy industry
D. her soils have been largely denuded of fertility

16. An island located at 0° latitude and 0° longitude would be located 16. B
A. exactly at the North Pole
B. where the Equator and the prime meridian meet
C. where the International Date Line and the Equator meet
D. exactly in the middle of the Pacific Ocean

17. An arrangement of mountain systems into large regional groupings forms a 17. A
A. cordillera B. mesa
C. range D. butte

18. All of the following cause poor soils in tropical forest lands EXCEPT 18. C
A. eluviation which causes a hard pan of clay to form below the surface
B. leaching which carries away minerals in solution
C. humus accumulation which leaves a thick layer of decomposed organic matter
D. migratory agriculture which leads to a succession of forest clearings, later abandoned to jungle growth

19. Which one of the following is NOT primarily known as a geographer? 19. D
A. George B. Cressey B. Norton Ginsburg
C. Preston James D. Jacob Viner

20. Which one of the following statements concerning manufacturing in New England is TRUE? 20. C
A. Its leading industries are the processing of foods and chemicals.
B. It no longer enjoys its former advantage of a highly skilled labor force.
C. Southern competition has virtually wiped out its once important textile industry.
D. Its industrial development has been based primarily upon an abundance of mineral reserves.

21. MOST of the Sahara Desert is a(n) 21. A
A. low plateau with a gravel or stony surface
B. endless expanse of sand dunes without vegetation
C. flat sandy surface resembling a beach, extending for a vast distance
D. mountainous area, interspersed with flat dusty plains

22. The Laurentian or Canadian Shield 22. C
   A. is a continuation of the Appalachian Chain
   B. extends across all of Canada
   C. is a much glaciated land of old, worn-down mountains
   D. is one of Canada's most important farming areas

23. Which one of the following geographic advantages has MOST furthered the development of iron and steel production in Birmingham, Alabama? 23. C
   A. The presence of a large and varied concentration of heavy industry nearby
   B. The high percentage of iron content in local ores
   C. The proximity of iron ore, coal, and limestone
   D. Cheap transportation of the Mesabi ores

24. The Corn Belt of the United States includes the 24. B
   A. states of the Appalachian Highlands
   B. area from central Ohio to Nebraska
   C. states of the Gulf Coastal Plain
   D. tier of states adjacent to the Canadian border

25. A geographic advantage that facilitated the early colonial settlement of Virginia was the 25. B
   A. protection afforded by a coastal mountain range
   B. navigability of rivers for ocean-going vessels
   C. extensive commercial fisheries in nearby coastal waters
   D. suitability of swiftly moving rivers and streams for water power

26. Which one of the following cities is a Baltic seaport? 26. A
   A. Gdansk B. Hamburg C. Leipzig D. Petsamo

27. A geographical feature of the African continent is that 27. B
   A. great rivers offer easy navigation far into the interior
   B. it has a regular coastline with few good harbors
   C. it is generally a lowland surrounded by coastal mountain ranges
   D. practically all of the continent is either tropical rainforest or desert

28. In which one of the following pairs is a city CORRECTLY matched with its climatic region? 28. D
   A. Nairobi, Kenya - Tropical rainforest
   B. Jakarta, Indonesia - Tropical savanna
   C. Paris, France - Humid continental
   D. Seattle, United States - Marine

29. Which one of the following statements is CORRECT? 29. B
   A. Each parallel of latitude is a Great Circle.
   B. Each meridian of longitude is one-half of a Great Circle.
   C. Latitude is measured from 0 to 180.
   D. Meridians of longitude are about 70 miles apart at all points on the globe.

30. The countries of Southern Europe in the Mediterranean climate belt have very few freezing days because 30. C
   A. these countries are near the Equator
   B. these countries are close to warming currents
   C. mountain barriers shut out most of the cold winds
   D. thick coastal forest covers absorb and retain large amounts of the sun's heat

31. Which one of the following is a geographic characteristic of South America? 31. B
   A. Its western cordillera has few minerals.
   B. Its coast line is fairly regular.
   C. Its largest area lies in the South Temperate Zone.
   D. Most of its good harbors are located on its west coast.

32. Which one of the following pairs of geographic factors BEST explains the economic importance of northern France? 32. A
   A. A network of navigable rivers and large coal deposits
   B. Mineral-laden mountains and heavy rainfall
   C. Large forests and extensive petroleum reserves
   D. Desirable climate and hydroelectric power

33. Under the provisions of the Antarctica Treaty that became effective in 1960, Antarctica 33. C
   A. is formally divided among the claimant nations
   B. assumes the status of a United Nations trusteeship
   C. is reserved for peaceful purposes with nuclear explosions forbidden
   D. is closed to further exploration and research in the interests of peace

34. Transportation in Russia has been aided MOST by 34. D
   A. its river systems
   B. its well-situated coastal areas
   C. the nearness of seaports to centers of population and production
   D. the general lack of natural barriers to hinder the movement of transportation

35. In comparison with the Southern Hemisphere, the Northern Hemisphere has 35. B
   A. fewer people
   B. more land area
   C. more water area
   D. less land area

36. If a plane traveling at 600 miles per hour left New York at 3 P.M. for San Francisco, at what hour, Pacific time, will it arrive? 36. C

   _____ P.M.

   A. 3   B. 4   C. 5   D. 7

37. Which one of the following statements BEST characterizes the Mercator projection? 37. D
A. There is a lack of distortion in the extreme latitudes.
B. Every straight line is an arc of a great circle on the globe.
C. Lines of latitude and longitude are curved.
D. It can be used to compute true direction.

38. A place located at 55 degrees north latitude and 60 degrees east longitude would be in 38. D
A. Canada B. India C. Norway D. Russia

39. Which one of the following rivers does NOT form part of an international boundary? 39. B
A. Rio Grande
B. Rhone
C. Yalu
D. Amur

40. The Piedmont is the area of the United States that 40. B
A. comprises the delta of the Mississippi River
B. lies between the Atlantic Coastal Plain and the Appalachian Mountains
C. constitutes the Central Valley of California
D. includes the wheat producing sections of the Great Plains

41. The ratio of the population of the United States to that of Canada is APPROXIMATELY 41. D
A. 2:1 B. 3:1 C. 5:1 D. 10:1

42. The United States ranks low in natural resources for 42. B
A. copper and zinc
B. chrome and tin
C. iron and lead
D. coal and tungsten

43. Which one of the following is TRUE of China? 43. B
A. Irrigation is rarely used because there is adequate rainfall in most farming areas.
B. A large part of the country has a monsoon climate.
C. The three major rivers flow mainly from north to south.
D. The British island of Hong Kong lies roughly midway between mainland China and Taiwan.

44. Which one of the following indicates the CORRECT order of passage in going from north to south in Russia? 44. A
A. Tundra-taiga-steppe
B. Steppe-taiga-tundra
C. Taiga-steppe-tundra
D. Tundra-steppe-taiga

45. Which one of the following groups includes the nations with the HIGHEST annual rate of population increase? 45. C
A. China, United Kingdom, Russia
B. China, India, Japan
C. Brazil, India, Indonesia
D. Brazil, Indonesia, West Germany

46. Which one of the following Southeast Asia nations has NO direct outlet to the sea? 46. B
A. Burma B. Laos
C. Vietnam D. Thailand

47. All of the following Mid-Eastern countries border on Turkmenistar and Azerbaijan EXCEPT 47. B
A. Iran B. Iraq
C. Turkey D. Afghanistan

48. *It has a shorter growing season, both from the standpoint of temperature and moisture, than any other cereal.* 48. C
A. Wheat B. Corn C. Barley D. Rice

49. Of the following, a nation which has a large deposit of zinc, silver, and lead ores is 49. C
A. India B. Union of South Africa
C. Mexico D. Brazil

50. A food introduced into Europe during the Crusades was 50. B
A. citrus fruits B. cane sugar
C. potatoes D. white bread

KEY (CORRECT ANSWERS)

1. A	11. A	21. A	31. B	41. D
2. A	12. A	22. C	32. A	42. B
3. A	13. A	23. C	33. C	43. B
4. D	14. D	24. B	34. D	44. A
5. D	15. C	25. B	35. B	45. C
6. B	16. B	26. A	36. C	46. B
7. C	17. A	27. B	37. D	47. B
8. C	18. C	28. D	38. D	48. C
9. C	19. D	29. B	39. B	49. C
10. B	20. C	30. C	40. B	50. B

TEST 2

DIRECTIONS: Each question or incomplete statement is followed by several suggested answers or completions. Select the one that BEST answers the question or completes the statement. *PRINT THE LETTER OF THE CORRECT ANSWER IN THE SPACE AT THE RIGHT.*

1. Submarine canyons cut deeply into the continental slope are now believed to have been formed by 1. A
   A. the action of turbidity currents
   B. stream erosion and subsequent submergence
   C. ground water solution
   D. submarine faulting

2. Jones Beach, Long Island, is located on a former 2. A
   A. offshore bar B. bay bar
   C. wave-cut bench D. coral reef

3. Long Island Sound was formed by *drowning* of a former 3. D
   A. geosyncline B. fault valley
   C. coastal plain D. inner lowland

4. New York State's Finger Lakes are examples of _____ lakes. 4. C
   A. cirque B. kettle
   C. moraine-dammed D. crater

5. The MAXIMUM size particle that a stream can carry is a measure of its 5. A
   A. competence B. capacity
   C. transporting power D. erosive index

6. Wind gaps are the result of 6. D
   A. wind action B. faulting
   C. glaciation D. stream piracy

7. Natural levees, alluvial fans, and braided stream channels are indications that 7. D
   A. load equals capacity B. load is less than capacity
   C. sea level is dropping D. load exceeds capacity

8. A watershed to which New York City has turned MOST recently to augment its water supply is that of the ____ River. 8. B
   A. Croton B. Delaware C. Hudson D. Susquehanna

9. A river flowing in a canyon it has made itself is in the stage called 9. A
   A. youth B. maturity
   C. late maturity D. old age

10. A glacier is said to be retreating when the ice 10. D
   A. reverses its direction of flow
   B. moves downhill faster than it melts at the terminus
   C. moves downhill at the same rate as it melts at the terminus
   D. moves downhill slower than it melts at the terminus

11. Lateral moraines of a valley glacier are PRINCIPALLY composed of rock debris 11. C
   A. deposited by glacial melt water
   B. freed from the enclosing glacial ice by frost-heaving
   C. derived from weathering of the valley walls
   D. which are well-sorted and stratified

12. Rocks are said to be metamorphosed if they undergo 12. D
   A. crystallization after melting
   B. intense alteration due to weathering
   C. compaction due to the weight of overlying sediments
   D. recrystallization without melting

13. A stream becomes permanent 13. A
   A. only after its bed has been cut below the water table
   B. only after it has become graded
   C. when it attains old age
   D. when it reaches base level

14. The FIRST streams to develop on an initial surface of erosion are called 14. A
   A. consequent
   B. subsequent
   C. obsequent
   D. antecedent

15. The BEST example of a volcano of the violently explosive type is 15. D
   A. Kilauea (Hawaii)
   B. Mt. Vesuvius (Italy)
   C. Lassen Peak (Calif.)
   D. Mt. Pelee (Martinique)

16. Canoe-shaped hills formed by glacial deposition are called 16. A
   A. drumlins
   B. eskers
   C. kames
   D. roches moutonnés

17. It would be unlikely to find ripple marks in sediments formed in (or by the action of) 17. A
   A. glaciers B. lakes C. rivers D. deserts

18. Graded bedding, examined from bottom to top, shows sediments ranging from 18. A
   A. gravel to sand to mud
   B. fine sand to coarse sand to mud
   C. mud to sand to gravel
   D. sand to gravel to mud

19. The type of rock MOST commonly found in batholiths is 19. C
A. rhyolite B. gabbro C. granite D. basalt

20. MOST earthquakes 20. B
A. are of the deep focus type
B. are the result of sudden earth movements
C. originate along the margins of the Atlantic Ocean basin
D. are associated with violent volcanic eruptions

21. The process of erosion is equivalent to 21. A
A. weathering followed by transportation
B. deposition followed by weathering
C. gradation followed by deposition
D. transportation followed by deposition

22. The name of the volcano which was formed in Mexico in 1943 is 22. C
A. Acapulco B. Orizaba C. Paricutin D. Toluca

23. The Appalachian Mountains were folded and uplifted to their GREATEST height at the close of the _____ Period. 23. B
A. Devonian B. Permian C. Triassic D. Tertiary

24. The MOST important single group of processes which serves to separate one time unit from another in the geologic time scale would be 24. A
A. diastrophism B. vulcanism
C. glaciation D. sedimentation

25. Of the following, the one that is a characteristic feature resulting from the re-elevation of the bed of an old river is that of 25. B
A. wider levees B. entrenched meanders
C. alluvial fans D. sandbars

26. If an airplane circumnavigates the earth 30 miles south of the North Pole, the TOTAL number of time zones over which it passes is which one of the following? 26. D
A. 3 B. 9 C. 12 D. 24

27. Assuming that the free end of a sand bar along the Atlantic coast extends first in a northwesterly direction, the current which produced it came 27. B
A. from the south
B. first from the south and then from the southeast
C. from the north
D. first from the north and then from the northwest

28. Of the following, the wind-blown sand that would produce the GREATEST angle of repose is _____ in size and _____. 28. A
A. small; angular B. small; rounded
C. large; angular D. large; rounded

29. The line along which a compass will indicate true north is called 29. B
A. true meridian
B. agonic
C. isochoric
D. isobaric

30. Of the following, it is TRUE that at the time of our winter solstice, the 30. D
A. south pole is in darkness for 24 hours
B. sun sets in the northwest
C. sun rises due east
D. sun's vertical rays are at the Tropic of Capricorn

31. The Great Salt Lake in Utah is a remnant of the earlier 31. A
A. Lake Bonneville
B. Lake Agassiz
C. Salton Sea
D. Lake Huron

32. The snow line is HIGHEST above sea level at which one of the following places? 32. B
A. Mt. Ranier, Washington
B. Mt. Kilimanjaro near the Equator
C. Mt. McKinley, Alaska
D. Mt. Vesuvius, Italy

33. Which one of the following would be the FIRST event in the cycle of the history of the development of a folded mountain chain? 33. A
A. Deposition in a geosynclinal trough
B. Isostatic adjustment
C. Synclinal folding
D. Block faulting

34. New Jersey is part of that physiographic region known as the 34. C
A. folded Appalachians
B. Appalachian Plateaus
C. Triassic Lowlands
D. Older Appalachians

35. The strike of a rock layer is measured with a compass in the field and found to be N45°E. What is its TRUE direction if the declination of the region is 12°W? 35. A
A. N33°E
B. N57°W
C. N57°E
D. S33°E

36. A landform sculptured out of gently dipping strata is a 36. C
A. butte
B. hogback
C. cuesta
D. mesa

37. Stalactites and stalagmites are formed by 37. A
A. loss of carbon dioxide from calcium bicarbonate solution
B. precipitation from super-saturated solutions in cavern pool
C. hot spring algae
D. differential erosion of cavern roofs and floors

38. The theory that a great sheet of ice had once covered northern Europe was FIRST advanced in 1840 by 38. A
A. Agassiz
B. deSaussure
C. Richter
D. Steno

39. Alluvial fans form when a river 39. B
A. enters a quiet body of water
B. emerges from a steep valley onto a flat area
C. plunges over a waterfall
D. emerges from a limestone cave

40. The competency of a stream is determined MOST readily by measuring which one of the following? 40. A
The
A. maximum size of its load
B. speed at which delta formation takes place
C. headwater erosion
D. formation of meanders

KEY (CORRECT ANSWERS)

1. A	11. C	21. A	31. A
2. A	12. D	22. C	32. B
3. D	13. A	23. B	33. A
4. C	14. A	24. A	34. C
5. A	15. D	25. B	35. A
6. D	16. A	26. D	36. C
7. D	17. A	27. B	37. A
8. B	18. A	28. A	38. A
9. A	19. C	29. B	39. B
10. D	20. B	30. D	40. A

TEST 3

DIRECTIONS: Each question or incomplete statement is followed by several suggested answers or completions. Select the one that BEST answers the question or completes the statement. *PRINT THE LETTER OF THE CORRECT ANSWER IN THE SPACE AT THE RIGHT.*

1. South America LACKS large and well-located supplies of 1. D
   A. water power B. oil C. nitrates
   D. coal E. copper

2. Large cities have NOT grown up along the Norwegian fiords because of 2. B
   A. poor natural harbors for ports
   B. the limited resources of the hinterland
   C. a long closed navigation season
   D. long winters with short days
   E. high tides and storms in harbors

3. The country with the world's GREATEST known reserves of petroleum is 3. A
   A. Russia B. Venezuela
   C. Saudi Arabia D. Kuwait
   E. the United States

4. The SHORTEST all-weather sea route between London and Tokyo is by way of the 4. C
   A. Northeast Passage B. Northwest Passage
   C. Suez Canal D. Cape of Good Hope
   E. Panama Canal

5. The area in the United States whose climate and products MOST closely approximate those of Egypt is 5. D
   A. the Gulf coast of Florida
   B. the Gulf coast of Texas
   C. Arizona
   D. the Imperial Valley of California
   E. the Salt Desert region of Utah

6. Which of these natural products has faced LEAST competition from recently developed synthetics? 6. B
   A. Rubber B. Tin C. Quinine
   D. Silk E. Chile saltpeter

7. On a globe, the tropics and the circles mark off zones of 7. C
   A. rainfall B. vegetation C. light
   D. air currents E. time

8. Pakistan is located _____ India. 8. E
   A. in northern
   B. in the central part of
   C. in the southern part of
   D. partly in the north and partly in the south of
   E. northwest of

9. The continent with the LARGEST area in the rainy tropics is 9. E
   A. Africa B. Asia C. Australia
   D. North America E. South America

10. Of the following countries, the one with the LARGEST population per square mile is 10. A
   A. Belgium B. China C. Egypt
   D. India E. Italy

11. The city of Africa that has the LARGEST population is 11. A
   A. Alexandria B. Cape Town C. Casablanca
   D. Johannesburg E. Tripoli

12. The population of the United States is APPROXIMATELY how many times that of Canada? 12. C
   _____ times.
   A. 3 B. 5 C. 10 D. 20 E. 30

13. Which of the following states produces practically no lumber? 13. D
   A. Arizona B. Idaho C. Kentucky
   D. Nevada E. Texas

14. In which of the Great Lakes does the water surface have LEAST elevation? 14. D
   A. Erie B. Huron C. Michigan
   D. Ontario E. Superior

Questions 15-23.

DIRECTIONS: Questions 15 through 23 are to be answered on the basis of the hypothetical map shown at the right.

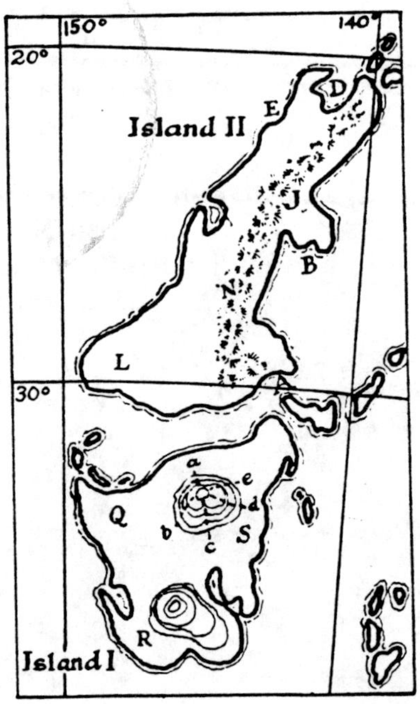

15. The latitudinal location of this island group is 15. B
A. N B. S C. E
D. W E. indeterminate

16. The longitudinal location of this island group is 16. D
A. N B. S C. E
D. W E. indeterminate

17. The ocean in which this island group would be located is the _____ Ocean. 17. E
A. Indian B. Arctic C. Antarctic
D. Atlantic E. Pacific

18. The group spreads over a north-south distance of about _____ miles. 18. C
A. 300 B. 690 C. 1300 D. 2000 E. 18

19. The approximate area of Island II is MOST NEARLY _____ square miles. 19. E
A. 5,000 B. 15,000 C. 25,000
D. 100,000 E. 150,000

20. If the prevailing winds are from the southeast, which of the following places probably has the LEAST rainfall? 20. D
A. B B. E C. J D. N E. L

21. On Island I, the STEEPEST slope on the hill is along path 21. A
A. a B. b C. c D. d E. e

22. Places that would experience the overhead sun at some time during the year are 22. C
A. L, Q B. R, S C. E, D D. A, B E. A, S

23. All of the following products might be grown on Island I EXCEPT 23. D
A. cotton B. corn C. sugar cane
D. rubber E. wheat

24. Factors that contributed to the early development of sailing among people living along the shores of the Mediterranean were: 24. C
I. High visibility of stars and landmarks in the dry summer air
II. Long periods of stormless weather
III. Many navigable rivers on which to gain experience
IV. Islands and peninsulas that provided easy stages for a journey

The CORRECT answer is:
A. I, II, III B. I, II, IV C. I, IV
D. II, III, IV E. III, IV

25. A GOOD natural harbor is characterized by: 25. B
I. Space for anchorage of vessels
II. Warehouses and storage facilities
III. Sufficient depth for large vessels
IV. Dredging equipment and dry docks
V. Protection from winds and storms

The CORRECT answer is:
A. I, II, III B. I, III, V C. I, III
D. II, V E. III, IV

26. Use of the southern pine forests in the United States has certain advantages in paper-pulp production over use of the northern spruce forests because 26. A
I. the long, moist open season favors growth
II. softwoods grow faster than hardwoods
III. the southern forests are nearer the market
IV. pulp can be made more easily by mechanical means in the South

The CORRECT answer is:
A. I *only* B. I, II C. I, IV
D. II, IV E. III, IV

27. Which of the following statements about Russia are true? 27. B
I. Murmansk may be reached via the Arctic Ocean.
II. There are more good river ports than seaports.
III. The Donets Basin is important as a grazing region.

The CORRECT answer is:
A. I *only* B. I, II C. I, III
D. II, III E. All of the above

28. The position of Great Britain as a commercial nation has been aided by 28. D
I. large supplies of petroleum within the British Isles
II. large supplies of coal within Great Britain
III. a large home iron and steel industry
IV. a surplus of agricultural products for export

The CORRECT answer is:
A. I, III B. I, III, IV C. I, IV
D. II, III E. II, III, IV

29. Factors that helped make the Ruhr one of the MAJOR industrial regions of the world are: 29. A
I. Rich iron ore deposits nearby
II. Large supplies of high quality coal within the region
III. Convenient inland water transportation
IV. Markets in Western Europe
V. Hydroelectric power from the Alps

The CORRECT answer is:
A. I, II, III, IV B. I, II, V C. I, III, IV
D. II, III, IV E. II, III, IV, V

30. Which of the following states appeared on the map of Europe for the FIRST time after World War I? 30. D
I. Bulgaria II. Czechoslovakia III. Poland
IV. Romania V. Yugoslavia

31 The CORRECT answer is:
A. I, II, IV, V B. II, III, IV C. II, III, V
D. II, V E. III, V

Questions 31-40.

DIRECTIONS: Questions 31 through 40 are to be answered on the basis of the following map. This map represents a hypothetical planet of the same size as the Earth and occupying the same relative position in a solar system identical with ours. It is supposedly governed by the same physical laws and experiences the same climatic conditions as the Earth.

31. An IMPORTANT advantage of this projection is that 31. E
A. shapes are not distorted
B. a single scale of miles may be used
C. it resembles most nearly the shape of the Earth
D. it is easier to read than other types
E. it shows the shortest route between any two places in the North Temperate Zone as an approximately straight line

32. Which of these countries probably has the WIDEST variation in climatic conditions? 32. B
A. 1 B. 6 C. 7 D. 9 E. 5

33. Which of these cities is farthest south? 33. A
A. A B. B C. C D. F E. L

34. What is the APPROXIMATE location of city D? 34. A
Longitude _____ latitude _____.
A. 150°W; 30°N B. 150°W; 30°S C. 30°N; 150°E
D. 30°N; 150°W E. 150°E; 30°S

35. A great circle route from city C to city M would pass APPROXIMATELY through 35. C
A. D B. G C. O
D. T E. the North Pole

36. A storm experienced at city Q would PROBABLY have been experienced earlier at 36. B
A. E B. G C. H
D. K E. the North Pole

37. The SHORTEST route between which two cities crosses the South Pole? 37. E
A. A and J B. B and M C. D and E
D. D and L E. J and K

38. In which direction is city D from city M? 38. B
A. North B. East C. West
D. Northeast E. Southwest

39. A plant placed in a window of a house in city K in June would receive MOST sunlight if the window faced 39. A
A. north B. east
C. south D. west
E. impossible to tell

40. Which of these cities is strategically located on the oceanic trade route from D to E? 40. E
A. G B. K C. M D. O E. T

KEY (CORRECT ANSWERS)

1. D	11. A	21. A	31. E
2. B	12. C	22. C	32. B
3. A	13. D	23. D	33. A
4. C	14. D	24. C	34. A
5. D	15. B	25. B	35. C
6. B	16. D	26. A	36. B
7. C	17. E	27. B	37. E
8. E	18. C	28. D	38. B
9. E	19. E	29. A	39. A
10. A	20. D	30. D	40. E

TEST 4

DIRECTIONS: Each question or incomplete statement is followed by several suggested answers or completions. Select the one that BEST answers the question or completes the statement. *PRINT THE LETTER OF THE CORRECT ANSWER IN THE SPACE AT THE RIGHT.*

1. Which of the following statements about the economic geography of the Far East are TRUE? 1. E
   I. The chief source of tin in that region is the Philippine Islands.
   II. New Zealand was our principal source of quinine before World War II.
   III. Petroleum refining in the Far East is concentrated in the Dutch East Indies.
   IV. Before this war, the Dutch East Indies produced about half the world's coffee supply.

   The CORRECT answer is:
   A. None of them B. I, II, IV C. II *only*
   D. II, III E. III *only*

2. Which of these statements about Brazil are TRUE? 2. C
   I. Brazil borders every South American country except Chile and Ecuador.
   II. It is the northernmost country of South America.
   III. Brazil is the only Portuguese-speaking country in South America.
   IV. Brazil is the world's chief source of rubber.

   The CORRECT answer is:
   A. I, II, III B. I, II, IV C. I, III
   D. II, IV E. All of the above

Questions 3-12.

DIRECTIONS: Questions 3 through 12 are to be answered on the basis of the map below.

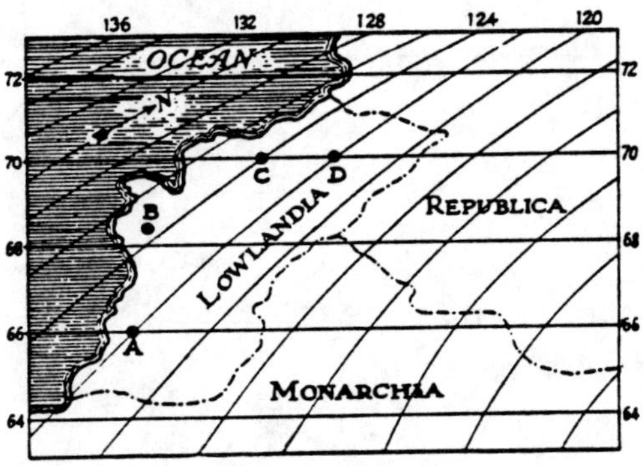

3\. The climate of Lowlandia probably approximates MOST closely that of 3. A
- A. tundras
- B. wet subtropical areas
- C. marine west coast
- D. rainy low latitudes
- E. continental deserts

4\. Assume that in this part of the world no attention is paid to standard time zones and that each city goes by apparent solar time - or sun time. 4. E
When it is 3:00 P.M. at A, what time is it at C?
_____ P.M.
- A. 2:44
- B. 2:52
- C. 3:00
- D. 3:08
- E. 3:16

5\. The distance between A and D is APPROXIMATELY _____ miles. 5. D
- A. 7
- B. 70
- C. 140
- D. 280
- E. 700

6\. Of the following occupations, which would MOST likely be found in this land? 6. D
- A. Cattle ranching
- B. Cotton growing
- C. Truck farming
- D. Reindeer herding
- E. Tulip growing

7\. Lowlandia is located _____ of the equator, _______ of Greenwich. 7. B
- A. north; east
- B. north; west
- C. north, due north
- D. south; east
- E. south; west

8\. Lowlandia is crossed by the 8. A
- A. Arctic Circle
- B. Equator
- C. Tropic of Capricorn
- D. Tropic of Cancer
- E. Antarctic Circle

9\. The day of the year on which the GREATEST number of hours of daylight are received at D will occur in 9. A
- A. June
- B. December
- C. March
- D. September
- E. April

10\. On the day of the year on which there is the GREATEST number of hours of daylight, how much daylight do people at D have? 10. E
______ hours.
- A. 4
- B. 8
- C. 12
- D. 16
- E. 24

11\. Traveling in a straight line from A to B, one would travel 11. E
- A. north
- B. northwest
- C. northeast
- D. slightly east of north
- E. slightly west of north

12\. If one were to travel due east from A by air, one would eventually reach 12. D
- A. Hawaii
- B. British Isles
- C. Australia
- D. Iceland
- E. Newfoundland

13. Which of the following are great circles? 13. D
I. Any circle drawn on the globe
II. Any circle bounding the zones
III. A circle drawn through the North Pole
IV. The equator
V. Any circle whose plane passes through the center of the earth

The CORRECT answer is:
A. I, IV, V
B. II, III
C. II, IV, V
D. IV, V
E. All of the above

14. A great circle is used in planning which of the following? 14. C
I. Ocean routes
II. Highways
III. International railroad routes
IV. Airplane routes

The CORRECT answer is:
A. I, II
B. I, III, IV
C. I, IV
D. IV *only*
E. All of the above

15. Of which of the following did the United States import MORE than 80% of its supply in 1970? 15. B
I. Tin
II. Copper
III. Rubber
IV. Coal
V. Oil
VI. Coffee
VII. Hemp

The CORRECT answer is:
A. I, II, III, V
B. I, III, VI, VII
C. II, III, IV, V, VI
D. IV, VI
E. All of the above

16. In summer, the monsoon winds of Southern Asia blow from the sea, whereas during the winter the winds blow from the land. 16. C
An IMPORTANT factor in this change is
A. the difference in the summer and winter distances of the earth from the sun
B. a reversal in the direction of the flow of ocean currents from winter to summer
C. the fact that the land gets warmer than the sea in summer and colder in winter
D. a marked seasonal variation in the relative humidity of sea air
E. a shift of the horse latitudes which move toward the south in summer and the north in winter

17. On a given July day in Bermuda, the moisture content of the air remains constant while the temperature rises. 17. D
What happens to the humidity in this situation?
A. The absolute humidity rises.
B. The absolute humidity decreases.
C. The relative humidity rises.
D. The relative humidity decreases.
E. Both the absolute humidity and the relative humidity remain constant.

18. The SHORTEST air route between two points is the 18. E
   A. great circle nearest to them
   B. great circle on which the point of departure lies
   C. great circle on which the point of destination lies
   D. greater arc of the great circle passing between them
   E. lesser arc of the great circle passing between them

19. The accompanying outline map represents 19. A
   A. Japan
   B. Java and Sumatra
   C. Borneo
   D. New Zealand
   E. the Philippines

20. 20. B

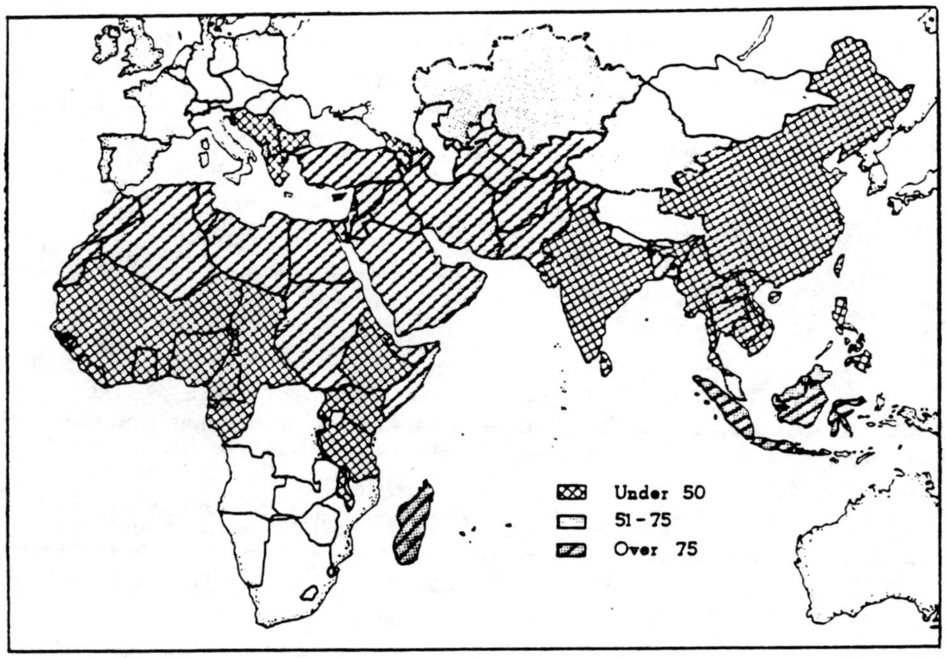

The above map was drawn to indicate
   A. population density per square mile
   B. percent of Muslims in total population
   C. annual mean temperature
   D. annual mean rainfall
   E. percent of arable land under cultivation

KEY (CORRECT ANSWERS)

1. E
2. C
3. A
4. E
5. D

6. D
7. B
8. A
9. A
10. E

11. E
12. D
13. D
14. C
15. B

16. C
17. D
18. E
19. A
20. B

EXAMINATION SECTION

TEST 1

DIRECTIONS: Each question or incomplete statement is followed by several suggested answers or completions. Select the one that BEST answers the question or completes the statement. *PRINT THE LETTER OF THE CORRECT ANSWER IN THE SPACE AT THE RIGHT.*

1. With which one of the following is the term aquifer MOST closely associated? 1. B
   A. Continental slope
   B. Artesian formation
   C. Mohole
   D. Rain clouds

2. With which one of the following are the four glacial periods MOST closely associated? 2. D
   A. Eocene
   B. Oligocene
   C. Miocene
   D. Pleistocene

3. The stage of a river characterized by a *V*-shaped valley is its 3. A
   A. youth
   B. early maturity
   C. late maturity
   D. old age

4. Of the following, recently uplifted shorelines are characterized by 4. D
   A. numerous bays
   B. estuaries
   C. peninsulas
   D. sandy beaches

5. The SHORTEST distance between New York City and Paris, France is along a(n) 5. D
   A. meridian
   B. parallel
   C. agonic line
   D. great circle

6. Of the following, which is an active volcano? 6. B
   A. Mt. Monadnock
   B. Lassen Peak
   C. Pike's Peak
   D. Mt. Washington

7. Tsunamis are produced by 7. A
   A. earthquakes in the ocean
   B. meteor showers
   C. glacial deposits
   D. erosion of limestone

8. Which one of the following regions shows GREATEST evidence of recent volcanic activity? 8. D
   A. Great Lakes
   B. Mississippi Valley
   C. East Coast
   D. Pacific Ranges

9. Areas of quiet water between the mainland and offshore bars are known as 9. D
   A. fiords
   B. tidal inlets
   C. estuaries
   D. lagoons

10. The type of moraine formed when an ice front is stationary for a relatively long period is 10. D
A. ground B. medial C. lateral D. terminal

11. Movement of the earth's solid crust is called 11. C
A. arête B. moraine
C. diastrophism D. cirque

12. The earth's crust is thinnest beneath which one of the following? 12. B
A. Coastal plains B. Oceans
C. Deserts D. Plateaus

13. Monadnocks are characteristic of which one of the following stages in the development of a mountain region? 13. D
A. Youth B. Early maturity
C. Late maturity D. Old age

14. An example among the following of a plateau formed by extrusive vulcanism is the 14. D
A. Appalachian B. Alleghany
C. Colorado D. Columbia

15. The south shore of Long Island is an example of which one of the following? 15. C
A. Terminal moraine B. Medial moraine
C. Outwash plain D. Drumlin

16. Of the following, the region that is CLOSEST to the earth is the 16. A
A. troposphere B. stratosphere
C. mesosphere D. ionosphere

17. The boundary between the polar winds and the prevailing westerlies of the temperate zones is the 17. A
A. polar front B. westerly front
C. polar edge D. polar circle

18. Approximately half of our atmosphere lies below an altitude (in feet) CLOSEST to 18. B
A. 7,000 B. 16,000 C. 30,000 D. 50,000

19. As compared to the average elevation of land surfaces, the average depth of the ocean is 19. B
A. unknown B. greater
C. less D. about the same

20. Of the following, the statement that is TRUE of sunspots is that 20. A
A. their temperature is about 4,000°C
B. they are found on all parts of the sun's surface
C. they reach a maximum every 11 months
D. they remain almost constant in size as they cross the sun's face

21. Waterfalls and rapids are characteristic of which one of the following stages in the erosional cycle of a stream? 21. A
   A. Youth B. Early maturity
   C. Late maturity D. Old age

22. An unconformity is MOST likely to be found in which one of the following? 22. C
   A. Lacustrine plain B. Seacoast plain
   C. Peneplane D. Glacial till plain

23. Of the following, the cause of Tsunamis is 23. C
   A. gravitation B. offshore currents
   C. crustal movements D. high winds

24. The Black Hills of South Dakota were formed PRIMARILY by 24. C
   A. crustal movements B. glacial erosion
   C. intrusive vulcanism D. extrusive vulcanism

25. A waterfall that has a resistant horizontal rock layer over less resistant layers will PROBABLY 25. A
   A. retreat upstream
   B. show a gradual decrease in height
   C. advance downstream
   D. form sinkholes

26. Of the following, a formation NOT associated with glaciation is a 26. D
   A. fiord B. hanging valley
   C. kettle lake D. plunge pool

27. An example, among the following, of an oozing or quiet volcano is 27. B
   A. Krakatoa B. Kilauea
   C. Mount Vesuvius D. Mount Lassen

28. A submerged shoreline is characterized by the presence of 28. D
   A. a very regular shoreline
   B. flat sandy land
   C. coastal plains
   D. many bays and inlets

29. Which one of these is a typical artesian structure? 29. D
   A. Clay resting on granite
   B. Sand resting on clay
   C. A bed of shale between beds of limestone
   D. A bed of sandstone between beds of shale

30. An IMPORTANT factor in the rapid recession of Niagara Falls is the 30. D
   A. columnar structure of a series of lava flows
   B. good jointing of a granite mass
   C. bed of soft shale overlying hard sandstones
   D. hard limestone overlying soft shales

31. A circular cavity ground out of bedrock by running water is called a 31. B
A. kettle hole
B. pothole
C. deflation hollow
D. blowhole

32. Explosive volcanoes USUALLY form _____ cones with _____ slopes. 32. B
A. cinder; gentle
B. cinder; steep
C. lava; steep
D. lava; gentle

33. The level, sandy southern half of Long Island is CHIEFLY a(n) 33. D
A. driftless area
B. ground moraine
C. lacustrine plain
D. outwash plain

34. Rivers flowing into a lake may eventually destroy it by 34. C
A. erosion
B. stream piracy
C. deposition
D. solution

35. The Hawaiian Islands differ from MOST other mountainous regions in that they 35. C
A. consist largely of metamorphic rock
B. were formed from large intrusions
C. have no geosynclinal history
D. consist largely of fossiliferous limestone

36. Among the following surface features, the one NOT formed by diatrophism is the 36. C
A. Colorado plateau
B. Sierra Nevada Mountains
C. Columbia plateau
D. Appalachian Mountains

37. The determination of temperatures of the ocean, during such times as the Devonian Period, are based upon chemical analysis of the fossil remains of that period, using as a clue the 37. A
A. ratio of oxygen-18 to carbon
B. ratio of radioactive carbon to oxygen
C. ratio of radioactive phosphorus to oxygen
D. quantity of radioactive calcium in the ocean

38. Of the following, the one which is believed to have been formed in the bed of a subglacial stream is 38. B
A. a drumlin
B. an esker
C. ground moraine
D. medial moraine

39. Of the following, the location with the HIGHEST snow line is 39. A
A. Mt. Kenya in Equatorial Africa
B. Pikes Peak in Colorado
C. Mt. Orizaba in Mexico
D. Mt. McKinley in Alaska

40. The BEST example of a quiet type of volcano, among the following, is 40. C
A. Krakatoa B. Lassen
C. Mauna Loa D. Vesuvius

41. Bays like San Francisco Bay, Chesapeake Bay, and Delaware Bay are 41. D
A. parts of deltas B. fiords
C. artificial harbors D. submerged river valleys

42. The Catskill Mountains of New York State can BEST be described as 42. C
A. coastal plain B. folded mountain
C. plateau D. a series of domes

43. A pair of rivers that marks the approximate southern limit of glaciation in North America during the Ice Age is the Missouri and the 43. B
A. Delaware B. Ohio
C. St. Lawrence D. Tennessee

44. The last great ice sheet which covered northern United States finally melted away APPROXIMATELY _____ years ago. 44. B
A. 1,000 B. 20,000 C. 200,000 D. 1,000,000

45. The ONLY active volcano in the fifty states, of the following, is 45. B
A. Katmai B. Lassen C. Shasta D. Paricutin

46. The Rossi-Forel scale is used to determine the 46. D
A. climatic conditions of the past
B. velocity of upper air movement
C. displacement of sediment by undersea currents
D. intensity of earthquakes

47. What is the longitude of a ship when its local time is 2 hours later than Greenwich time? 47. C
A. 150°E B. 10°E C. 30°E D. 30°W

48. The time meridian of the Mountain Standard Time Belt in the United States is 48. B
A. 95°W B. 105°W C. 110°W D. 120°W

49. At apparent solar noon, the shadow of a flagpole in New York City points true north 49. A
A. on any day of the year B. only in winter
C. only in summer D. only on the equinoxes

50. A ship sailing eastward crosses the International Date Line at 50. B
A. 7:02 P.M. Friday B. 7:02 P.M. Wednesday
C. 8:02 P.M. Wednesday D. 8:02 P.M. Thursday

KEY (CORRECT ANSWERS)

1. B
2. D
3. A
4. D
5. D
6. B
7. A
8. D
9. D
10. D
11. C
12. B
13. D
14. D
15. C
16. A
17. A
18. B
19. B
20. A
21. A
22. C
23. C
24. C
25. A
26. D
27. B
28. D
29. D
30. D
31. B
32. B
33. D
34. C
35. C
36. C
37. A
38. B
39. A
40. C
41. D
42. C
43. B
44. B
45. B
46. D
47. C
48. B
49. A
50. B

TEST 2

DIRECTIONS: Each question or incomplete statement is followed by several suggested answers or completions. Select the one that BEST answers the question or completes the statement. *PRINT THE LETTER OF THE CORRECT ANSWER IN THE SPACE AT THE RIGHT.*

1. Rivers flowing in steep-sided valleys are classified as which one of the following? 1. B
   A. Drowned B. Young C. Old D. Mature

2. If the tide is highest at 3 P.M., it will be lowest at a time CLOSEST to which one of the following? 2. C
   A. 3 A.M. B. 6 P.M. C. 9 P.M. D. 12 Midnight

3. Artesian wells are successful where an aquifer lies between layers of 3. B
   A. pumice B. shale C. sand D. sandstone

4. The Yosemite Valley in California is noted for which one of the following? 4. A
   A. Hanging valleys
   B. Geysers
   C. Lakes
   D. Natural bridges

5. The snow line is HIGHEST in which one of the following? 5. B
   A. Maine B. Mexico C. Iceland D. New York

6. Headward erosion by valley glaciers results in the formation of which one of the following? 6. B
   A. Drumlins B. Cirques C. Kames D. Eskers

7. Salt lakes are found in which one of the following regions? 7. A
   A. Great Basin
   B. Great Plains
   C. Allegheny Plateau
   D. Appalachian Mountains

8. When it is 11 P.M. standard time, Thursday, in Los Angeles, the day and standard time in New York City is 8. D
   A. 8 P.M. Thursday
   B. 12 Midnight Friday
   C. 1 A.M. Friday
   D. 2 A.M. Friday

9. An oxbow lake is MOST closely associated with which one of the following? 9. C
   A
   A. captured stream
   B. glacier
   C. meandering stream
   D. water gap

10. Of the following, the condition that characterizes a shoreline that has been recently uplifted is that of 10. D
    A. bays and inlets
    B. drowned valleys
    C. sea caves
    D. sandy beaches

11. Sinkholes result from which one of the following actions? 11. C
A. Land submergence
B. Whirlpool erosion
C. Cavern roof collapse
D. Melting of embedded ice

12. The sandy deposit on a sidewalk next to a sloping lawn after a rainfall is MOST closely analogous to which one of the following? 12. D
A. Stalagmite
B. Delta
C. Crevasse
D. Alluvial fan

13. A peneplane is very CLOSELY associated with which one of the following? 13. C
A. Volcanic activity
B. Oxbox lakes
C. Monadnocks
D. Young mountains

14. It is believed that all major earthquakes are the result of 14. B
A. volcanic activity
B. faulting
C. underground cave-ins
D. landslides

15. The GREATEST amount of insolation is received from those rays of the sun that strike the earth's surface at an angle of 15. C
A. 0°
B. 45°
C. 90°
D. 120°

16. Tides are caused PRINCIPALLY by 16. A
A. gravitational attraction of sun and moon
B. the wind
C. periodic changes in atmospheric pressure
D. diastrophic forces within the earth

17. When the winds are prevailing westerlies, a desert tends to form 17. B
A. west of a high mountain range
B. east of a high mountain range
C. independently of a high mountain range
D. on the tops of the mountains

18. An igneous intrusion which forms dome mountains is known as a 18. C
A. sill
B. batholith
C. laccolith
D. dike

19. The standard time meridian for the Central Standard Time Belt is 19. B
A. 75°W
B. 90°W
C. 105°W
D. 120°W

20. Of the following conditions, the one which HINDERS the preservation of fossil specimens is 20. B
A. dry or cold climate
B. decay of dead organisms
C. presence of hard portions in organisms
D. speedy burial of dead organisms

21. When it is 4 P.M. July 4th at 105° west longitude, the time and date at 150° east longitude is 21. C
A. 9 P.M., July 4
B. 9 A.M., July 6
C. 9 A.M., July 5
D. 9 P.M., July 5

22. If a watch is used to find the direction of north, the hour hand should be pointed 22. C
A. due east
B. due west
C. at the sun
D. at the zenith

23. Which one of the following inventions has made the chronometer relatively unimportant in the determination of longitude? 23. A
A. Radio
B. Sonar
C. Loran
D. Radar

24. Spring tides and neap tides occur at a given port 24. B
A. at the same time
B. about one week apart
C. about two weeks apart
D. about 12 hours apart

25. Which one of the following statements about the sun in New York City on December 21st is TRUE? 25. B
A. Rises north of east
B. Sets south of west
C. *Rides* high in the sky
D. Rises at due east

26. Contour plowing can help all of the following EXCEPT 26. D
A. sheet erosion
B. rill erosion
C. wind erosion
D. gullying

27. Tyrannosaurus rex roamed the continents during which one of the following periods? 27. A
A. Cretaceous
B. Permian
C. Devonian
D. Cambrian

28. The earth's crust is THINNEST beneath 28. C
A. deserts
B. lacustrine plains
C. oceans
D. plateaus

29. More than half (by weight) of our atmosphere is found in the zone of the 29. D
A. exosphere
B. ionosphere
C. stratosphere
D. troposphere

30. The Palisades of the Hudson River is the face of a 30. C
A. dike
B. laccolith
C. sill
D. sedimentary formation

31. Of the following, the BEST explanation for the formation of oxbow lakes is the 31. B
A. persistence of lagoons in areas of shoreline emergence
B. straightening out of an old stream, thereby leaving remnants of former meanders
C. collection of water from underground streams as they reach the earth's surface
D. collection of water from melting glaciers

32. The celestial equator at New York City is located 32. C
A. at an angle of 23.5° to the earth's Equator
B. at an angle of 41° to the earth's Equator
C. in the plane of the earth's Equator
D. in the plane of the ecliptic

33. In MOST cases, foci of earthquakes are located 33. B
A. within a depth of one mile of the earth's surface
B. between the earth's surface and a depth of 450 miles
C. anywhere from 450 to 1,800 miles below the earth's surface
D. anywhere from 1,800 to 3,160 miles below the earth's surface

34. Of four identical buildings, the one LEAST likely to suffer damage from earthquakes is the one built on 34. D
A. old lake sediments
B. glacial till
C. deep residual mantle
D. bedrock

35. The zone of the atmosphere that is CLOSEST to the earth's surface is called the 35. D
A. exosphere
B. ionosphere
C. stratosphere
D. troposphere

36. Which one of the following functions served by the oceans is LEAST important to man at the present time? 36. D
A. Regulating climate
B. A means of travel and shipping
C. A source of moisture
D. A source of minerals

37. Of the following, the one which is a feature of the ocean floor is a 37. C
A. continental shelf which is no more than four or five miles wide from Newfoundland to Massachusetts
B. mid-Atlantic ridge which separates the North Atlantic Ocean from the South Atlantic Ocean
C. number of trenches each of which may exceed 1,000 miles in length and reach a depth of approximately 6 or 7 miles
D. series of seamounts which are more numerous in the Atlantic Ocean than in the Pacific Ocean

38. The deposits on the ocean floor at a depth greater than 15,000 feet consist LARGELY of 38. C
A. sediment derived from the land
B. shells of dead sea organisms
C. volcanic and meteoric dust
D. calcareous ooze

39. A ship whose local time is 2 hours and 30 minutes earlier than its chromometer time is at which one of the following longitudes? 39. B
A. 20°E
B. 35°W
C. 47°W
D. 75°W

40. If it is 10:30 P.M. standard time Thursday at the 150°E meridian, the standard time and day at the 150°W meridian is 40. B
A. 2:30 A.M. Wednesday B. 2:30 A.M. Thursday
C. 6:30 A.M. Thursday D. 2:30 A.M. Friday

41. Of the following types of areas, the one in which earthquakes can be expected to occur MOST frequently is an area with 41. C
A. peneplanes B. many mountain glaciers
C. young mountains D. dissected plateaus

42. All of the following are formed by deposition EXCEPT 42. B
A. alluvial fans B. monadnocks
C. valley trains D. volcanic cones

43. Which one of the following methods is used to determine mean sea level? 43. A
A. Finding the midway point between high water and low water
B. Using the low water mark only
C. Using the high water mark only
D. Measuring the ocean depth at a known point

44. Lateral moraines of a valley glacier are PRINCIPALLY composed of rock debris 44. C
A. deposited by glacial melt water
B. freed from the enclosing ice by frostheaving
C. derived from the weathering of the valley walls
D. which are well sorted and stratified

45. The STEEPEST slopes of volcanic cones are formed by 45. D
A. viscous lava
B. fluid lava
C. alternating lava flows and cinder beds
D. pyroclastic material

46. Truncated, cone-shaped mountains, called guyots, which have broad platforms 3,000 to 6,000 feet below sea level and are located principally in the Pacific Ocean, are believed to be 46. A
A. wave-cut volcanic peaks which later subsided
B. remnants of former continents
C. peaks of the Mid-Pacific Ridge
D. igneous intrusions exposed by ocean currents

47. A recently proposed explanation of the origin of submarine canyons, also believed to explain a break in the trans-Atlantic cable in 1929, is the existence of 47. D
A. faulting
B. submarine volcanic explosions
C. internal wave oscillation between water masses of different density
D. turbidity currents

48. Reelfoot Lake in Tennessee is a famous example of a(n) 48. D
A. oxbow lake
B. lake resulting from vulcanism
C. lake resulting from a raft blockade
D. lake resulting from an earthquake

49. Lake Ronkonkoma on Long Island is an example of a lake known as 49. D
A. playa B. cirque C. finger D. kettle

50. The Palisades of New Jersey is an example of a 50. C
A. dike B. batholith C. sill D. monadnock

KEY (CORRECT ANSWERS)

1. B	11. C	21. C	31. B	41. C
2. C	12. D	22. C	32. C	42. B
3. B	13. C	23. A	33. B	43. A
4. A	14. B	24. B	34. D	44. C
5. B	15. C	25. B	35. D	45. D
6. B	16. A	26. D	36. D	46. A
7. A	17. B	27. A	37. C	47. D
8. D	18. C	28. C	38. C	48. D
9. C	19. B	29. D	39. B	49. D
10. D	20. B	30. C	40. B	50. C

TEST 3

DIRECTIONS: Each question or incomplete statement is followed by several suggested answers or completions. Select the one that BEST answers the question or completes the statement. *PRINT THE LETTER OF THE CORRECT ANSWER IN THE SPACE AT THE RIGHT.*

1. A butte is a landform developed by 1. B
   A. glacial erosion in a mountainous region
   B. sub-aerial erosion in a region of horizontal strata
   C. differential erosion of steeply dipping strata
   D. wave action along a rocky coast

2. The development of a stream is said to enter the stage of maturity when 2. D
   A. its capacity exceeds its load
   B. it has developed rapids and waterfalls along its course
   C. it has built natural levees along its banks
   D. it has established a longitudinal profile of equilibrium

3. The velocity of a stream is LEAST affected by 3. B
   A. shape of its channel
   B. composition of the rocks over which it flows
   C. amount of load it carries
   D. gradient of its bed

4. The effectiveness of headward erosion of a stream depends, among other factors, upon the 4. C
   A. volume of water in the stream at its mouth
   B. length of the stream course
   C. difference in elevation between the source and the mouth of the stream
   D. amount of material deposited by the stream in the form of a delta

5. Ground water accomplishes its geologic work CHIEFLY by 5. B
   A. abrasion
   B. solution
   C. attrition
   D. hydraulic action

6. Alluvial fans tend to form where 6. A
   A. rivers emerge from mountain canyons
   B. streams enter a quiet body of water
   C. drifting sand develops into sand dunes
   D. glacial meltwater flows out of a terminal moraine

7. The glacier that covered New York State during the Ice Age originated in 7. D
   A. Greenland B. Siberia C. Alaska D. Labrador

8. The Catskills are designated as — 8. A
   A. a dissected plateau B. folded mountains
   C. a coastal plain D. domed mountains

9. Features resulting from wave erosion of the underlying rocks along a rocky coast are called — 9. C
   A. fringing reefs B. promontories
   C. sea cliffs D. estuaries

10. The direction in which the Ice Sheet traveled over New York State may be MOST directly determined by an examination of — 10. C
   A. kames B. kettle holes
   C. drumlins D. potholes

11. Loess is a deposit formed CHIEFLY by — 11. B
   A. valley glaciers B. wind
   C. rivers D. ground water

12. Oxbow lakes are evidence of — 12. C
   A. glacial action B. stream piracy
   C. mature rivers D. down-cutting

13. Which one of the following terms describes a stream which has maintained its course across an uplift which it antedates? — 13. D
   A. Subsequent B. Consequent
   C. Superimposed D. Antecedent

14. Ventifacts may be defined as — 14. C
   A. columns of air moving upward in convection systems
   B. shallow basins created by wind erosion
   C. wind faceted rocks, sometimes called dreikanters
   D. desert pavements resulting from deflation

15. Which one of the following resulted in the phosphate deposits in Utah and Montana? — 15. C
   A. Volcanic action
   B. Weathering of igneous rocks
   C. Marine deposits
   D. Folding and faulting

16. The number of degrees south latitude at which the vertical rays of the noonday sun are located on November 23 is CLOSEST to which one of the following? — 16. C
   A. 10 B. 12 C. 16 D. 20

17. According to the theory of isostasy, the density of — 17. C
   A. land masses is greater than that of ocean basins
   B. mountain masses is greater than that of surrounding plains
   C. land masses is smaller than that of ocean basins
   D. land masses is the same as that of ocean basins

18. At zone time 11:00 P.M. on June 5, a ship is at longitude 179°30' West on course West; four hours later, the ship is at longitude 179°30' East. The zone time and date at the latter instant are 18. D
A. 3 A.M. June 6
B. 3 A.M. June 5
C. 3 P.M. June 6
D. 3 A.M. June 7

19. A wave cut cliff located thirty feet above sea level along the California coast line is evidence of 19. C
A. submergence
B. earthquake activity
C. uplift
D. glacial erosion

20. The term *saltation* will MOST likely be used in the discussion of the 20. A
A. movements of particles at the bottom of a stream
B. evidence of age of earth by studying composition of ocean water
C. evidence of former submergence of land by sea
D. securing minerals in tropical climates

21. If the strike of an inclined tabular rock is N45°E, its dip will PROBABLY be 21. B
A. E W
B. S E
C. N E
D. S N

22. The constituent which comprises over 60% of the gases discharged by fumaroles is 22. D
A. oxygen
B. sulfur dioxide
C. carbon dioxide
D. steam

23. Of the following, the BEST description of tarn is that it is a 23. C
A. valley whose floor is notably higher than the floor of the valley into which it leads
B. sticky, gray colored clay, the product of chemically weathered till
C. small lake formed in the bottom of a cirque when the glacier melts
D. large hole carved in the bottom of a falls by the concurrent action of water and rocks

24. Seismologists are able to compute the distance between an earthquake epicenter and the recording station by determining the difference between the arrival times of the waves indicated in which one of the following pairs? 24. A
A. Primary and secondary
B. Primary and tertiary
C. Secondary and tertiary
D. Tertiary and quaternary

25. The angle of inclination or declination of bedrock to the horizontal is known as 25. A
A. dip
B. strike
C. fault angle
D. gradient departure

Questions 26-32.

DIRECTIONS: Questions 26 through 32 are to be answered on the basis of the following diagram.

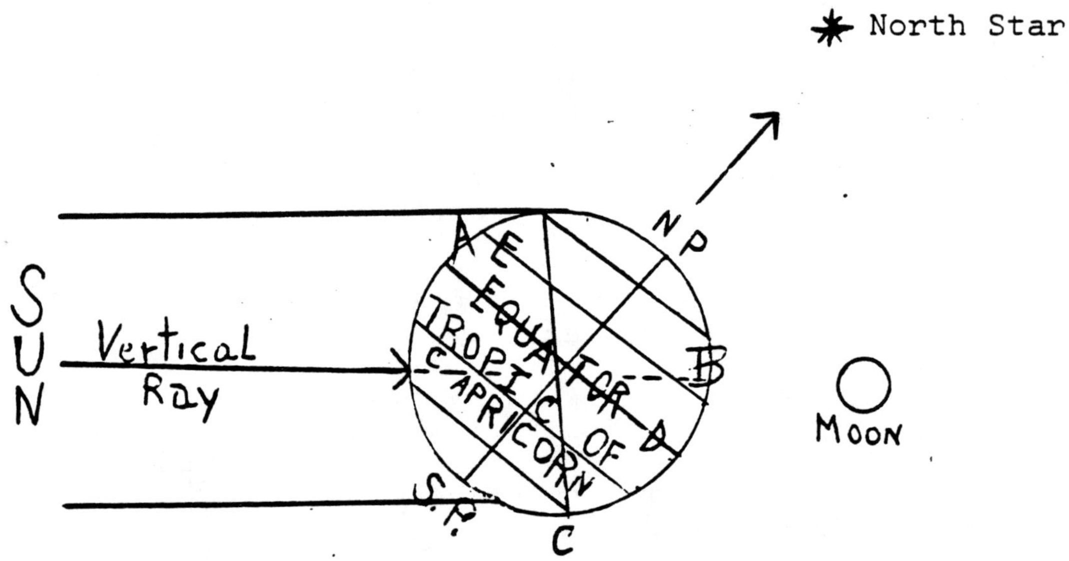

26. Which one of the following is the latitude of point E? 26. B
A. 18°N B. 23½°N
C. 66½°N D. None of the above

27. Since A is on the 90° West meridian, the solar time along the prime meridian is 27. D
A. 6 A.M. B. 12 P.M. C. 12 A.M. D. 6 P.M.

28. A is on the 90° West meridian; therefore, B is located on the 28. A
A. 90° East meridian B. 90° West meridian
C. 105° East meridian D. none of the above

29. The date at which the sun will next set at the south pole is CLOSEST to 29. B
A. September 23 B. March 21
C. June 21 D. December 21

30. If you were standing at a point on the Arctic Circle, the number of degrees by which you would see the North Star above the horizon would be which one of the following? 30. D
A. 0 B. 23½ C. 50 D. 66½

31. Which one of the following BEST expresses the number of degrees that the noon sun is above the Equator? 31. C
A. 0 B. 23½ C. 66½ D. 90

32. Assuming that the moon is NOT in eclipse, which one of the following phases of the moon is shown by the diagram? 32. C
A. New gibbous
B. New
C. Full
D. Last quarter

33. The thickness of the earth's crust and its general structure can be inferred from a study of 33. D
A. radioactive dating of sedimentary rocks
B. the rate of submergence and emergence of shorelines
C. the variation in depths of the world's oceans
D. seismic waves, both natural and manmade

34. Faulting of the earth's crust may produce 34. A
A. block mountains
B. synclines and anticlines
C. laccoliths
D. monadnocks

35. An irregular surface that separates two series of rocks of different geological ages is a(n) 35. D
A. terrace
B. paleozoic remnant
C. old regolith
D. unconformity

36. The Black Hills of South Dakota are an example of the type of mountains called 36. B
A. block B. dome C. fold D. volcanic

37. Of the following, the feature MOST closely associated with a young river is a(n) 37. A
A. V-shaped valley
B. meandering course
C. oxbow lake
D. broad flood plain

38. Of the following, which one is MOST characteristic of the Permian Period? 38. B
A. The inland basins of eastern United States covered by shallow seas
B. Extensive deposits of rock salt and gypsum forming through Kansas, Oklahoma, and Texas
C. The rapid growth of ferns, giant rushes, and newly developed conifers
D. The recent ice sheets still covering Labrador and Northern Canada

39. Debris which collects at the bottom of a cliff is called a(n) 39. C
A. sand bar B. monadnock C. talus D. esker

40. An upfold of rock layers is called a(n) 40. A
A. anticline B. intrusion C. syncline D. drumlin

41. Geologists can estimate the end of the Ice Age in the United States by counting or measuring the 41. A
A. bands in varved clays
B. schistose bands in scoured bedrock
C. chi-square distribution of erratics
D. amount of radioactivity of bedrock in Northern Canada

42. Of the following, which one is a remnant of the ancient lake, Bonneville? 42. B
   A. Lake Superior B. Great Salt Lake
   C. The Dead Sea D. Salton Sea

43. From which one of the following can the direction in which a glacier traveled be determined? 43. B
   A. The shapes of kettles
   B. Striae on bedrock
   C. The slope of the land
   D. Direction of river flow

44. Tsunamis are waves that may be caused by 44. D
   A. offshore currents
   B. gravitational effects of the moon and sun
   C. strong, continuous winds blowing in one direction
   D. diastrophism

45. The term *water table* refers to 45. D
   A. the impervious layer below an aquifer
   B. a shallow lake
   C. a mirage
   D. the level below which ground water is found

46. The part of any glaciated valley that has been flooded by the sea is called a 46. C
   A. lagoon B. bay C. fjord D. tarn

47. All of the following are typical features of alpine topography EXCEPT the 47. C
   A. arête B. cirque C. doline D. tarn

48. Which one of the following is the BEST means of preventing the flooding of a river valley? 48. B
   A. Building levees on the river banks
   B. Damming the headwaters of tributary streams
   C. Planting natural vegetation along the banks
   D. Constructing revettments

49. A horst is the opposite of which one of the following? 49. A
   A. Graben B. Hogback
   C. Synclinal ridge D. Fault scarp

50. Which one of the following is characteristic of a shoreline of submergence? 50. A
   A. Numerous bays and inlets
   B. Long stretches of smooth, sandy beaches
   C. Shallow marshy areas
   D. Offshore bars

KEY (CORRECT ANSWERS)

1. B	11. B	21. B	31. C	41. A
2. D	12. C	22. D	32. C	42. B
3. B	13. D	23. C	33. D	43. B
4. C	14. C	24. A	34. A	44. D
5. B	15. C	25. A	35. D	45. D
6. A	16. C	26. B	36. B	46. C
7. D	17. C	27. D	37. A	47. C
8. A	18. D	28. A	38. B	48. B
9. C	19. C	29. B	39. C	49. A
10. C	20. A	30. D	40. A	50. A

TEST 4

DIRECTIONS: Each question or incomplete statement is followed by several suggested answers or completions. Select the one that BEST answers the question or completes the statement. *PRINT THE LETTER OF THE CORRECT ANSWER IN THE SPACE AT THE RIGHT.*

1. The Aleutian, Kurile, and Hawaiian Islands are the result of 1. D
   A. differential erosion of a former mainland
   B. submergence of mountain chains
   C. faulting of a former continental margin
   D. volcanic activity

2. Trilobites were MOST abundant in the _____ Era. 2. C
   A. Cenozoic B. Mesozoic
   C. Paleozoic D. Proterozoic

3. Of the following, the CLOSEST precursors of amphibians are 3. B
   A. ostracoderms B. crossopterygians
   C. sharks D. placoderms

4. Niagara Falls goes over an escarpment formed of rocks of the following period: 4. D
   A. Cambrian B. Devonian C. Triassic D. Silurian

5. Of the following, an animal once believed to be extinct but since found to be living is the 5. A
   A. coelacanth B. brachiopod
   C. belemnite D. ostracod

6. The MOST widespread transgression of Paleozoic seas upon the continent of North America occurred in which one of the following periods? 6. B
   A. Cambrian B. Ordovician
   C. Pennsylvanian D. Silurian

7. The maximum age, in years, of ancient remains for the dating of which the C^{14} method can be used, is CLOSEST to which one of the following? 7. C
   A. 5,000 B. 10,000 C. 40,000 D. 100,000

8. The prehistoric creature stegocephalian was a(n) 8. B
   A. fish B. amphibian C. reptile D. mammal

9. In the earlier period of oil exploration and discovery, MOST of the oil was found in which one of the following? 9. B
   A. Salt domes B. Anticlines
   C. Unconformities D. Ancient shore lines

10. An outwash plain is a(n) _____ fluvial deposit. 10. C
A. stratified B. unstratified
C. stratified glacio- D. unstratified glacio-

11. Which one of the following animals became extinct EARLIEST in geologic time? 11. C
A. Trilobites B. Dinosaurs
C. Graptolites D. Ammonites

12. Which one of the following orogenies in the geologic history of North America is MOST ancient? 12. B
A. Acadian B. Taconian
C. Laramide D. Appalachian

13. A radial drainage pattern is developed in which one of the following types of regions? 13. D
A. Plateaus B. Fault block mountains
C. Folded mountains D. Volcanic mountains

14. Knob and kettle topography is COMMONLY developed on 14. D
A. sandy deserts B. coastal plains
C. outwash plains D. terminal moraines

15. The youthful stage of a shoreline of emergency TYPICALLY has 15. D
A. headlands B. tombolos
C. wave-cut terraces D. lagoons

16. Which one of the following supported the doctrine of catastrophism? 16. C
A. Agassiz B. Lyell C. Cuvier D. Hutton

17. The continued erosion of a mesa may result in the formation of 17. A
A. butte B. cuesta C. flatiron D. hogback

18. The competency of a stream refers to the 18. B
A. ability of the stream to carry all the sediment dropped into it
B. largest sized particles that the stream may carry
C. ability of the stream to cross resistant bands of rock
D. ability of the stream to erode its bed

19. Which one of the following BEST describes typical offshore bars? 19. B
A. Both the seaward and the landward sides are straight.
B. The landward side is irregular and the seaward side is straight.
C. The landward side is straight and the seaward side is irregular.
D. Both the seaward and the landward sides are irregular.

20. A stream which flows along the axis of an anticlinal valley is classified as a(n) _____ stream. 20. B
   A. consequent B. subsequent
   C. obsequent D. superimposed

21. Which one of the following mountains in the United States can be considered an active volcano? 21. B
   A. Mt. Shasta B. Mt. Katmai
   C. Mt. Rainier D. Mt. Hood

22. Which one of the following is an example of a lava plateau? 22. C
   _____ Plateau.
   A. Allegheny B. Colorado
   C. Columbia D. Piedmont

23. The estuary of the Hudson River is PRINCIPALLY the result of 23. B
   A. deposition of silt at its mouth
   B. subsidence of the coast line
   C. tidal bores
   D. glacial scouring

24. Fine examples of shield volcanoes may be found in 24. A
   A. the Hawaiian Islands B. the West Indian Islands
   C. Japan D. Mexico

25. Yosemite National Park is famous CHIEFLY for its 25. B
   A. geysers and hotsprings B. deeply glaciated valleys
   C. underground caverns D. mesas and buttes

26. Newark, New Jersey lies in the physiographic province known as the 26. C
   A. New England Upland B. Piedmont
   C. Triassic Lowland D. Coastal Plain

27. In recent years, vast deposits of high-grade iron ore have been discovered in 27. C
   A. Mexico B. Alaska C. Labrador D. Colorado

28. Of the following mountain ranges, the one in which glaciation is still an effective erosional agent is the _____ Mountains. 28. C
   A. Adirondack B. Green
   C. Rocky D. White

29. Of the following, the PRINCIPAL agent of erosion is 29. A
   A. running water B. ground water
   C. the atmosphere D. glacial ice

30. Development of cross-bedding is LEAST likely to occur in sedimentary deposits laid down by 30. B
   A. wind B. ice
   C. ocean currents D. streams

31. Using eastern North America and the adjacent Atlantic Ocean basin as an example, the surface showing the LEAST relief is the 31. B
   A. continental platform B. continental shelf
   C. continental slope D. ocean basin proper

32. Dinosaurs were abundant during the geologic period called the 32. A
   A. Jurassic B. Eocene
   C. Mississippian D. Pennsylvanian

33. The diameter of the earth at the Equator is greater than that measured between the poles because of the earth's 33. A
   A. rotation around its axis
   B. revolution around the sun
   C. magnetic field
   D. gravitative attraction by the sun

34. A young river is characterized by 34. D
   A. deltas B. gentle slopes
   C. low velocity D. waterfalls

35. Scientific deductions concerning probable conditions existing in the earth's core are LEAST reliable when dealing with its 35. A
   A. temperature B. pressure
   C. composition D. density

36. Which one of the following deposits accumulated as a result of mass wasting? 36. A
   A. Talus B. Volcanic ash
   C. Loess D. Alluvium

37. A fold whose limbs are inclined in the same general direction but at different angles of dip can be BEST described as 37. D
   A. open B. isoclinal
   C. symmetrical D. overturned

38. In every 100 pounds of ordinary sea water, the weight of dissolved NaCl is APPROXIMATELY _____ pounds. 38. B
   A. 2 B. 2.75 C. 3.5 D. 4.25

39. Natural gas from the gas wells of petroleum regions consists PRINCIPALLY of 39. D
   A. acetylene B. carbon monoxide
   C. hydrogen D. methane

40. With the exception of nitrogen and oxygen, the MOST abundant gas in the air near the earth's surface is 40. A
   A. argon B. carbon dioxide
   C. helium D. neon

41. Sub-Saharan Africa may MOST correctly be described as featuring 41. D
A. rivers that are easily navigable and that offer access to inland areas
B. broad, productive coastal plains with a wide offshore continental shelf
C. a coastline that offers many deep and sheltered harbors
D. a great plateau with fringing mountains and a generally narrow coastal plain

42. Which of the following terms describes a climate characterized by warm wet winters and hot dry summers? 42. D
A. Humid continental B. Humid subtropical
C. Marine D. Mediterranean

43. In which of the following pairs do both countries border on China? 43. B
A. Burma and Thailand B. India and Tibet
C. Laos and Afghanistan D. Pakistan and Cambodia

44. The MAJORITY of the people in the Republic of Indonesia are 44. D
A. Christians B. Hindus
C. Buddhists D. Moslems

45. Which of the following CORRECTLY associates a geographer with his idea? 45. A
A. Ellsworth Huntington - Climatic determinism is a factor in civilization
B. Halford Mackinder - Russia's location limits the role it can play in world affairs
C. Alfred Thayer Mahan - Control of land masses is the decisive force in world politics
D. Ellen Semple - A nation's history is determined by the ideas and culture of its people

46. Which of the following map projections provides an image of the earth's surface MOST similar to that provided by a globe? 46. C
A. Azimuthal B. Mercator
C. Orthographic D. Conic

47. The region in the eastern slopes of the Andes which is composed MAINLY of tropical rainforest with some grassland is called the 47. D
A. pampas B. latifundia
C. llanos D. montana

48. Considering Eurasia as a single continent, it may MOST correctly be stated that 48. B
A. the line of the Urals and the straits separates marked physical differences
B. the great desert and steppe areas begin east of Turkey and the Urals

C. the Asian portion of Eurasia has the longest coastline in proportion to its total land area
D. in the Asian portion plains and low elevations are more predominant than in Europe

49. Which of the following is a TRUE statement about geographic conditions in Russia? 49. A
A. Parts of the Black Sea coast receive almost 100 inches of rainfall annually.
B. The severity of the climate is moderated by the influence of the bordering oceans.
C. Summer temperatures along the Arctic coast rarely exceed 32° Fahrenheit.
D. Over 40% of the country is mountainous and exceeds 6,560 feet in elevation.

50. A geographic feature common to the Union of South Africa, Liberia, and Ethiopia is that they are all located 50. D
A. south of the Equator
B. in tropical rainforests
C. on coastal plains
D. on plateaus

KEY (CORRECT ANSWERS)

1. D	11. C	21. B	31. B	41. D
2. C	12. B	22. C	32. A	42. D
3. B	13. D	23. B	33. A	43. B
4. D	14. D	24. A	34. D	44. D
5. A	15. D	25. B	35. A	45. A
6. B	16. C	26. C	36. A	46. C
7. C	17. A	27. C	37. D	47. D
8. B	18. B	28. C	38. B	48. B
9. B	19. B	29. A	39. D	49. A
10. C	20. B	30. B	40. A	50. D

TEST 5

DIRECTIONS: Each question or incomplete statement is followed by several suggested answers or completions. Select the one that BEST answers the question or completes the statement. *PRINT THE LETTER OF THE CORRECT ANSWER IN THE SPACE AT THE RIGHT.*

1. The United States relies ALMOST ENTIRELY on imports for its supply of 1. C
   A. copper B. titanium C. nickel D. zinc

2. The Hudson River at New York City is a(n) 2. B
   A. strait B. estuary
   C. fjord D. mature stream

3. The Niagara River flows from Lake _____ into _____. 3. A
   A. Erie; Lake Ontario
   B. Ontario; Lake Erie
   C. Erie; the St. Lawrence River
   D. Ontario; the St. Lawrence River

4. The International Geophysical Year 4. D
   A. is the first cooperative, international venture for scientific exploration
   B. was organized under the auspices of the United Nations
   C. was conducted by a Committee of official representatives of the governments of all participating nations
   D. officially lasted from July 1, 1957 to December 31, 1958

5. If the axis of the earth were inclined 10° instead of $23\frac{1}{2}$°, summers in New York City would be 5. A
   A. cooler B. warmer C. longer D. shorter

6. The weight of an object on the earth's surface is NOT affected by 6. A
   A. air pressure
   B. altitude
   C. the earth's rotation
   D. local variations in rock density

7. The aims of the TVA in harnessing the Tennessee River include all of the following EXCEPT 7. D
   A. prevention of erosion B. control of floods
   C. recreation D. control of rainfall

8. An outstanding characteristic of a young river is 8. D
   A. a low gradient B. a meandering course
   C. muddy water D. waterfalls

9. Of the following mountain ranges, the one estimated to be the YOUNGEST is
   A. Alps
   B. Rockies
   C. Adirondacks
   D. Himalayas

9. B

10. About 100,000,000 years ago, North America
   A. had approximately the same form as today except that a land bridge connected Alaska and Siberia
   B. was connected by land to Asia, South America, and Europe
   C. was divided into two parts by a great inland sea
   D. was connected by a land bridge to Africa

10. C

11. Of the following epochs, the one which is NOT associated with the Mesozoic Era is
   A. Cretaceous
   B. Jurassic
   C. Triassic
   D. Permian

11. D

12. Large-scale ocean currents are caused by
   A. erosion
   B. winds
   C. boat traffic
   D. tidal waves

12. B

13. Coney Island is an example of which one of the following? A(n)
   A. barrier beach
   B. atoll
   C. levee
   D. spit

13. A

14. The fraction of the solid part of the earth that lies beneath water and ice is
   A. 1/3
   B. ½
   C. 3/4
   D. 5/6

14. C

15. The surface of the solid part of the earth consists essentially of huge blocks of rock which sometimes break away and rise or sink.
This action is called
   A. volcanic action
   B. folding
   C. faulting
   D. weathering

15. C

16. The new volcano which was formed in Mexico in 1943 is named
   A. Popocatepetl
   B. Mauna Loa
   C. Paricutin
   D. Krakatoa

16. C

17. A narrow V-shaped valley is considered by geologists to be
   A. young
   B. old
   C. mature
   D. between mature and old

17. A

18. As compared with a valley made by a stream, a glacial valley is
   A. rounded and smooth
   B. irregular and rough
   C. irregular and smooth
   D. rounded and rough

18. A

19. The branch of geology dealing with the origin of land forms is called 19. A
   A. geomorphology B. geopolitics
   C. geomology D. geotology

20. The farthest south in what is now the continental United States of America that the last glacier reached was 20. C
   A. Maine B. New York C. Kentucky D. Alabama

21. The expressions *hanging wall* and *foot wall* are used in reference to 21. D
   A. top and bottom of a sedimentary stratum
   B. the two limbs of a fold
   C. top and bottom of an erosional escarpment
   D. the two sides of a fault

22. Lava which solidifies under water is USUALLY which of the following structures? 22. B
   A. Cindery B. Pillow C. Ropy D. Vesicular

23. In order for lava flows to exhibit pillow structure, they must have 23. C
   A. been highly viscous when molten
   B. contained large quantities of dissolved gases
   C. flowed into water or over wet ground
   D. erupted explosively

24. A dendritic drainage pattern is developed in a region of 24. C
   A. folded mountains B. volcanic mountains
   C. plateaus and plains D. cuestas and hogbacks

25. A feature found in sedimentary rocks which CANNOT be used to distinguish the top and the bottom of beds is 25. B
   A. delta cross-bedding B. current ripple marks
   C. graded bedding D. fossil mudcracks

26. Of the following processes, the one CHIEFLY responsible for the formation of coal is 26. A
   A. distillation B. decomposition
   C. petrifaction D. replacement

27. Hanging troughs are formed by the differential action of 27. C
   A. weathering B. stream erosion
   C. valley glaciation D. continental glaciation

28. A drumlin is a(n) _____ deposit. 28. D
   A. stratified stream B. unstratified stream
   C. stratified glacial D. unstratified glacial

29. Spits are deposits laid down by 29. C
   A. streams entering the ocean
   B. action of shallow-water waves
   C. longshore currents
   D. wind blowing along the shore

30. The FIRST streams which develop on an initial land surface are classified as 30. C
A. subsequent B. superimposed
C. consequent D. obsequent

31. Which one of the following statements about a stream while it is in its stage of youthful development is CORRECT? Its 31. A
A. capacity exceeds its load
B. load exceeds its capacity
C. capacity and load are independent variables
D. load equals its capacity

32. The taiga of the northern regions of North America and Eurasia is characterized by 32. B
A. deciduous forests B. coniferous forests
C. treeless plains D. cold deserts

33. The Llanos of Colombia, the Campos of Brazil, and the Sudan in Africa are examples of 33. C
A. deserts B. forests C. savannas D. jungles

34. The equatorial belt of calms is called the 34. A
A. doldrums B. tropical drift
C. aurora D. baguio

35. The warm, dry wind which blows down the eastern slopes of the Rocky Mountains and over the adjacent plains in the United States and Canada is called 35. B
A. sirocco B. chinook C. bora D. mistral

36. In general, major ocean currents move through the waters of the earth in a(n) 36. A
A. clockwise direction in the Northern Hemisphere and in a counter-clockwise direction in the Southern Hemisphere
B. clockwise direction both north and south of the Equator
C. true circle north of the Equator and in straight lines south of the Equator
D. easterly direction in the Northern Hemisphere and in a westerly direction in the Southern Hemisphere

37. In normal years, the SMALLEST variation in mean temperature throughout the years occurs in 37. A
A. Singapore B. Calcutta
C. Baghdad D. Naples

Questions 38-39.

DIRECTIONS: Questions 38 and 39 are to be answered on the basis of the graph below.

CITY X

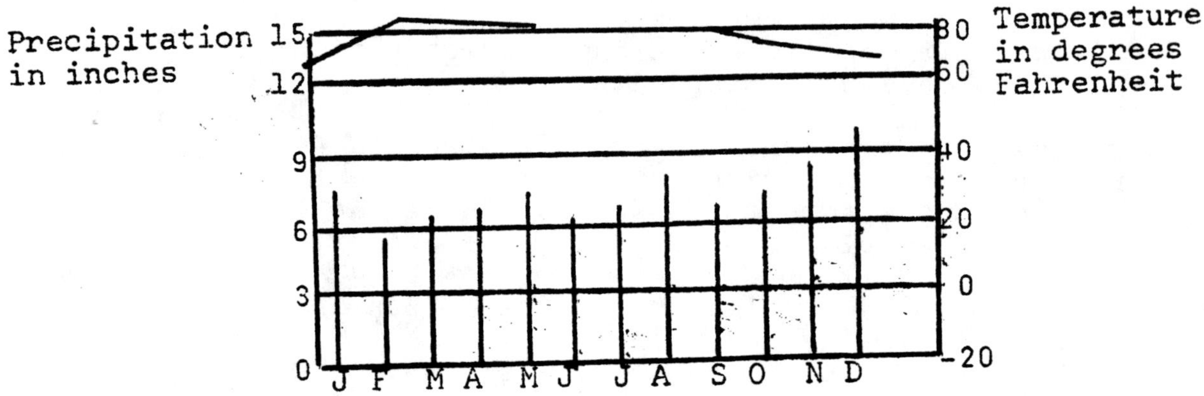

38. The climate of city X may BEST be characterized as 38. D
   A. humid continental
   B. temperate marine
   C. monsoon tropical
   D. wet tropical

39. The crops which might be grown in the hinterland of city X are 39. B
   A. corn, wheat, rice
   B. cacao, rubber, coconuts
   C. peaches, apples, cherries
   D. citrus fruits, flax, rye

40. A geographic feature common to the cities of Shanghai, Rangoon, Karachi, and Calcutta is that they are located 40. C
   A. between 20° and 25° north latitude
   B. on lake plains
   C. on or near delta plains
   D. on Piedmont plateaus

41. In which one of the following groups are all three cities in APPROXIMATELY the same latitude? 41. D
   A. Chicago, Moscow, Oslo
   B. Valparaiso, Rio de Janeiro, Capetown
   C. Tokyo, New York, London
   D. Philadelphia, Madrid, Ankara

42. The city named below which is MOST NEARLY the same distance south of the Equator as New York City is north of the Equator is 42. D
   A. Saigon, South Vietnam
   B. Sao Paulo, Brazil
   C. Port Elizabeth, Republic of South Africa
   D. Wellington, New Zealand

43. The climate of the British Isles is MOST NEARLY like that of the coastal areas of 43. B
A. Newfoundland
B. Washington and Oregon
C. New England
D. Western Australia

44. The LARGEST producer of oil in the Middle East is 44. A
A. Saudi Arabia
B. Kuwait
C. Iran
D. Iraq

45. For which of the following reasons has the Arctic become MORE important in recent years? 45. B
I. The Great Circle airline route
II. The development of mineral resources
III. The establishment of meteorological stations
IV. The exploitation of food resources

The CORRECT answer is:
A. I, II B. I, III C. I, IV D. I, II, III

46. The CHIEF disadvantage of the Mercator Projection is that 46. B
A. it is not useful for navigation
B. the size of land areas is seriously distorted at some latitudes
C. the shapes of land areas are distorted to maintain a constant ratio between the sizes of areas on the map and those on the earth's surface
D. it is always polar centered and, therefore, very difficult to understand

47. The Catskill Mountains of New York State can BEST be described as 47. C
A. coastal plain
B. folded mountain
C. plateau
D. a series of domes

48. Bays like San Francisco Bay, Chesapeake Bay, and Delaware Bay are 48. D
A. parts of deltas
B. fiords
C. artificial harbors
D. submerged river valleys

49. A very meandering course, flat topography with no falls or rapids, poor drainage on the flood plain, swampy vegetation, and few tributaries are CHARACTERISTIC of a river in its 49. C
A. youth
B. maturity
C. old age
D. maturity and old age

50. Potholes, outwash plains, hanging valleys, and kames are all terms DIRECTLY associated with the activities of 50. B
A. rivers B. glaciers C. volcanoes D. mountains

KEY (CORRECT ANSWERS)

1. C	11. D	21. D	31. A	41. D
2. B	12. B	22. B	32. B	42. D
3. A	13. A	23. C	33. C	43. B
4. D	14. C	24. C	34. A	44. A
5. A	15. C	25. B	35. B	45. B
6. A	16. C	26. A	36. A	46. B
7. D	17. A	27. C	37. A	47. C
8. D	18. A	28. D	38. D	48. D
9. B	19. A	29. C	39. B	49. C
10. C	20. C	30. C	40. C	50. E

EXAMINATION SECTION

TEST 1

DIRECTIONS: Each question or incomplete statement is followed by several suggested answers or completions. Select the one that BEST answers the question or completes the statement. *PRINT THE LETTER OF THE CORRECT ANSWER IN THE SPACE AT THE RIGHT.*

1. All of the following have made significant contributions to the development of geography teaching EXCEPT 1. B
   A. Isaiah Bowman B. Henry C. Lea
   C. Ellen C. Semple D. J. Russell Smith

2. Which one of the following groups consists ENTIRELY of minerals for which the U.S. is today dependent almost entirely on imports from foreign countries? 2. D
   A. Lead, sulphur, tungsten
   B. Copper, graphite, mercury
   C. Zinc, manganese, asbestos (strategic)
   D. Tin, nickel, diamonds (industrial)

3. Which one of the following is NOT characteristic of regions of cyclonic storms? 3. D
   A. Rain falls in moderate abundance at all seasons.
   B. Ellsworth Huntington says they are the most important parts of the world.
   C. Weather is subject to marked changes every few days.
   D. Seasons are not strongly pronounced.

4. In which one of the following pairs of industries is the location of the first influenced largely by proximity to raw materials and of the second largely by proximity to fuel or markets? 4. C
   A. Canning - paper manufacturing
   B. Machinery manufacturing - cotton ginning
   C. Paper manufacturing - machinery manufacturing
   D. Glass-making - sugar refining

5. The areas of greatest industrialization throughout the world correspond MOST closely with 5. A
   A. the quality and accessibility of coal deposits
   B. natural soil fertility
   C. the quantity of potential hydroelectric power
   D. seasonal uniformity of climatic conditions

6. The Strait of Malacca is significant for all of the following EXCEPT that 6. A
   A. it is the passageway between Formosa and the mainland of China
   B. routes from Europe to Australia cross here
   C. it is strategically located where the Pacific and Indian Oceans converge
   D. the Great Circle air routes from the North American West Coast to the Orient find their termini here

7. During the past quarter century, the GREATEST degree of competition has existed in the production of 7. A
A. oil B. aluminum C. tin D. nickel

8. Trade among European countries resembles trade among our own states in that 8. B
A. water traffic plays the same relatively important role in both regions
B. each region has manufacturing areas concentrated in the districts of middle temperatures facing the Atlantic
C. similar types of grain constitute the bulk of the traffic in foodstuffs
D. the southern route has better natural approaches to the hinterland than the northern route

9. The discovery of high grade iron ore in Venezuela 9. C
A. constituted the first important industrial raw material supplied by that country to the United States
B. added to the great surpluses of high grade iron ore now stored in the United States
C. served to compensate for the serious depletion suffered by the Mesabi range during and since World War II
D. increased the tonnage passing through the Panama Canal

10. The relatively high standard of living of Sweden, Switzerland, Denmark, and the Netherlands prevails in spite of the fact that each country 10. A
A. lacks either coal or iron
B. lacks both coal and iron
C. remained neutral in both World Wars
D. barely experienced the industrial revolution

11. Strategic location has been a key factor in explaining the international economic importance in ancient, medieval and modern times, respectively, of 11. A
A. Athens, Venice, London B. Cairo, Florence, London
C. Rome, Madrid, Liverpool D. Carthage, London, Milan

12. An economically destructive result of the use of chemical fertilizers has been the fact that it 12. A
A. permits farming to continue on unfertile land which should be allowed to become wooded to stop erosion
B. requires excessive quantities of electricity in the manufacture of potash
C. has encouraged the wasteful slaughter of animals for ground bone as a source of phosphates
D. deprives the steel industry of necessary supplies of phosphorus

13. Which one of the following grassland areas has the characteristic short, shallow-rooted grasses of the steppe? 13. C
   A. Central United States, particularly in Illinois, Iowa, and Oklahoma
   B. The Argentine Pampa
   C. Regions west of the Rockies, particularly in California, Oregon, and Washington
   D. Plains of the Danube in Hungary and Rumania

14. Which one of the following factors is NOT responsible for the highly fertile loessial soils? 14. C
   A. Unweathered minerals
   B. Porous structure
   C. Coarse texture
   D. High water-holding capacity

15. The MOST characteristic cause of the formation of the karst plain is 15. C
   A. the work of glaciers
   B. wind erosion
   C. limestone erosion
   D. stream deposition

16. A food consumption map of the world would show that, as density of population increases, 16. A
   A. the consumption of meat declines
   B. wheat replaces rice as the staff of life
   C. number of calories consumed daily per capita increases
   D. millet becomes the chief source of nourishment

17. In the case of which of the following products did World War II effect *little* change in the organization of primary production but *great* change in processing? 17. B
   A. Natural rubber
   B. Ornamental diamonds
   C. Copper
   D. Chemical fertilizers

18. Which one of the following areas is MOST likely to have the greatest tidal range? 18. D
   A. A small island in mid-ocean, such as Wake Island
   B. A nearly enclosed body of water, such as the Baltic Sea
   C. Sheltered water, such as the Gulf of Mexico
   D. Funnel-shaped bays or estuaries, such as Liverpool harbor

19. The growing importance of the Gulf ports of the United States reflects the fact that 19. C
   A. they draw the bulk of their freight from beyond the Ohio and Missouri Rivers
   B. more of their freight is going to Europe and less is passing through the Panama Canal
   C. the South is making rapid strides toward a more balanced economy
   D. the decline in locally produced raw materials has caused them to seek freight from outside the region

20. In which of the following ways do the Great Lakes and the railroads complement each other as traffic arteries? 20. A
I. The termini of most of the railroads of the middle west are on the Great Lakes.
II. The world's greatest railway center is at the juncture of three of the Great Lakes.
III. Railroads carry high class freight between the same cities served by lake steamers for low cost bulky freight.
IV. Railroads carry all of the freight during winter months when the lakes are frozen.

The CORRECT answer is:
A. I, III, IV
B. I, III
C. I, II, IV
D. I, II, III, IV

21. The map projection which gives the HIGHEST degree of distortion in the polar areas is the 21. A
A. Mercator
B. Lambert
C. Molleweide
D. Gnomonic

22. A diagram of the wind belts of the world would show that the U.S. was in the region of the 22. A
A. Prevailing Westerlies
B. Horse Latitudes
C. Northeast Trade Winds
D. Doldrums

23. The term *shelter belt* is usually associated with 23. A
A. conservation
B. horticulture
C. irrigation
D. lumbering

24. The Don Basin is known as the 24. D
A. chief gold mining district of Russia
B. Riviera coast of Russia
C. terminal of the northern sea route
D. industrial center of southern Russia

25. Which of the following are characteristics of a Mediterranean climate? 25. A
I. Most of the rainfall is in the winter.
II. There is a drought, more or less complete, in the summer.
III. There is a low percentage of sunshine in winter.
IV. The winters are mild.

The CORRECT answer is:
A. I, II, IV
B. I, II, III
C. I, III, IV
D. II, III, IV

KEY (CORRECT ANSWERS)

1. B	6. A	11. A	16. A	21. A
2. D	7. A	12. A	17. B	22. A
3. D	8. B	13. C	18. D	23. A
4. C	9. C	14. C	19. C	24. D
5. A	10. A	15. C	20. A	25. A

TEST 2

DIRECTIONS: Each question or incomplete statement is followed by several suggested answers or completions. Select the one that BEST answers the question or completes the statement. *PRINT THE LETTER OF THE CORRECT ANSWER IN THE SPACE AT THE RIGHT.*

1. In which of the following pairs of rivers is the first used PRIMARILY for hydroelectric power and the second PRIMARILY for navigation? 1. A
   A. Tennessee and Danube
   B. Colorado and Zambezi
   C. Hudson and Tennessee
   D. Danube and Rhine

2. Which one of the following pairs CORRECTLY groups a natural resource with an area in which large quantities of that natural resource are found? 2. A
   A. Copper - Andes Mountains, Chile
   B. Chromium - Lorraine, France
   C. Petroleum - Katanga, Belgian Congo
   D. Asbestos - Sudbury, Canada

3. In which one of the following areas does the presence of coal, limestone, and iron in close proximity account for its industrial development? 3. D
   A. Le Creusot, France
   B. Essen, Germany
   C. Manchester, Great Britain
   D. Birmingham, Alabama

4. Which one of the following pairs includes natural resources which can be secured by digging wells? 4. C
   A. Mercury, chromium
   B. Oil, potash
   C. Salt, sulphur
   D. Asbestos, natural gas

5. The present location of atomic energy plants is the U.S. illustrates the importance of their proximity to 5. C
   A. labor markets
   B. sources of raw materials
   C. cheap sources of power
   D. sparsely settled areas

6. Which statement is NOT true of a North Polar projection map? 6. C
   A. It exaggerates the shape of areas south of the Equator.
   B. It shows parallels of latitude as concentric circles.
   C. It is the most useful type of map for ship navigators.
   D. The shortest distance between any two points on the map is a great circle.

7. From Montreal, ships can get to Duluth via the Great Lakes and 7. B
   A. Welland and Erie Canals
   B. Welland and Sault Ste. Marie Canals
   C. Erie and Sault Ste. Marie Canals
   D. Sault Ste. Marie Canal and Hudson River

8. The Volga River flows into the 8. A
   A. Caspian Sea B. Black Sea
   C. Arctic Ocean D. Mediterranean Sea

9. Tropical savanna climate is characteristic of all of the following countries EXCEPT 9. C
   A. India B. Thailand C. Arabia D. Sudan

10. Which of the following has the GREATEST influence on the climate of Europe? 10. A
   A. Prevailing westerlies B. Doldrums
   C. Monsoon D. Trade winds

11. Which of the following cities is a leading center of jute mills? 11. B
   A. Bombay B. Calcutta C. Shanghai D. Manila

12. Manganese is an important product in all of the following areas EXCEPT 12. A
   A. Canada B. Russia
   C. South Africa D. India

13. Which of the following countries is NOT correctly paired with one of its important products? 13. D
   A. Chile - copper B. Bolivia - tin
   C. Peru - petroleum D. Argentina - coal

14. To which of these climate types does MOST of western Europe belong? 14. B
   A. Continental B. Marine
   C. Temperate D. Mediterranean

15. Which of the following is NOT an important asset of Sweden? 15. C
   A. Iron B. Timber
   C. Coal D. Water power

16. Domestic production of which of the following comes closest to meeting Japan's needs? 16. C
   A. Aluminum B. Coal C. Copper D. Gold

17. The outstanding iron ore producing area in the U.S. is the 17. B
   A. Birmingham, Alabama area B. Lake Superior region
   C. Pennsylvania area D. Rocky Mountain area

18. The only one of the Great Lakes WHOLLY within the U.S. is 18. A
   A. Michigan B. Superior C. Erie D. Huron

19. The LARGEST petroleum producing area in the United States is the _____ field. 19. B
   A. Ohio-Indiana B. Mid-Continent
   C. Gulf D. California

20. The Canadian port which is ice-bound for part of the year is — 20. C

A. Halifax
B. St. John's
C. Montreal
D. Vancouver

21. The length of a degree of latitude is about _____ miles. — 21. C

A. 43
B. 56
C. 69
D. 82

22. On which of the following map projections are parallels NOT shown as horizontal lines? — 22. C

A. Homolographic
B. Interrupted homolosine
C. Polyconic
D. Sinusoidal

23. Among the important metals are — 23. A

I. tin
II. manganese
III. copper
IV. nickel

When these are listed according to the degree to which the United States is self-sufficient in them, the PROPER order is:

A. III, II, IV, I
B. III, IV, II, I
D. II, IV, III, I
D. IV, III, I, II

24. Which of the following are CORRECTLY associated with the Mediterranean type of agriculture? — 24. A

I. Emphasis on tree and bush crops
II. Irrigated winter crops
III. Abundance of livestock
IV. Largest acreage in cereals

The CORRECT answer is:

A. I, IV
B. I, II
C, II, IV
D. II, III

25. Which of the following statements concerning the geography and economy of Australia are ACCURATE? — 25. D

I. Agriculture is the chief occupation of most of the settled region of the continent.
II. The Murray-Darling river system drains the west slopes of the Great Dividing Range.
III. The chief climatic handicap is deficient and unreliable rainfall.
IV. Iron and coal are available as a base for industrial development.

The CORRECT answer is:

A. I, III, IV
B. II, IV
C. I, IV
D. II, III, IV

KEY (CORRECT ANSWERS)

1. A	6. C	11. B	16. C	21. C
2. A	7. B	12. A	17. B	22. C
3. D	8. A	13. D	18. A	23. A
4. C	9. C	14. B	19. B	24. A
5. C	10. A	15. C	20. C	25. D

TEST 3

DIRECTIONS: Each question or incomplete statement is followed by several suggested answers or completions. Select the one that BEST answers the question or completes the statement. *PRINT THE LETTER OF THE CORRECT ANSWER IN THE SPACE AT THE RIGHT.*

1. Which of the following regions of the world are examples of the SAME climate type? 1. A
   A. Central India, Northern Zimbabwe, Brazilian Highlands
   B. Manchuria, southeastern Canada, Norway
   C. Southern China, southeastern United States, western Australia
   D. Great Plains of U.S., western China, central Europe

2. Which of the following crops is INCORRECTLY associated with a climatic condition which permits its growth? 2. D
   A. Wheat - climate too dry for the production of corn
   B. Millet - summers too cool or dry for efficient rice production
   C. Rye - summers too cool for the production of corn
   D. Corn - climate too dry for the production of wheat

3. Which of the following is NOT a detrimental effect of irrigation? 3. C
   A. Increased soil salinity
   B. Waterlogging
   C. Decrease in alkaline material on surface of soil
   D. Change in character of the soil

4. The natural grasslands of North America included MOST of the present states of 4. B
   A. Wyoming, Colorado, Utah
   B. Illinois, Iowa, Nebraska
   C. Ohio, Indiana, Illinois
   D. Michigan, Indiana, Kentucky

5. The monsoons are LESS marked in North America than in other continents because of the influence of the 5. D
   A. arid regions in the western part of the United States
   B. ocean currents off the Atlantic and Pacific coasts
   C. Rocky and Appalachian mountain systems
   D. Great Lakes and Hudson Bay

6. Which of the following are CORRECTLY associated with land utilization on the Great Plains of the United States? 6. B
   I. Commercial grazing of sheep
   II. Commercial grazing of cattle
   III. Wheat farming
   IV. Production of forage crops in irrigable districts
   V. Production of sugar beets in irrigable districts

The CORRECT answer is:
A. I, II, IV
B. I, II, III, IV, V
C. II, III, IV, V
D. II, IV

7. The FUNDAMENTAL reason for the difference in climate between eastern United States and western Europe lies in 7. D
A. their latitude
B. their topography
C. the direction of the prevailing winds
D. the side of the continent on which they are located

8. Which of the following statements concerning the geography and economy of Israel are ACCURATE? 8. D
I. The mountain ranges are of vital importance to irrigation.
II. The projected developed of the Jordan valley contemplates use of Jordan River water for irrigation.
III. Haifa is the terminus of the main pipeline from the Mosul oil fields.
IV. Wheat is an important crop but cereals and flour are imported.

The CORRECT answer is:
A. I, III, IV
B. II, III
C. I, IV
D. I, II, III, IV

9. There is little livestock in China today because 9. D
A. during World War II the Japanese removed most cattle to Japan
B. the *cultural revolution* resulted in the depletion or destruction of China's livestock
C. the Chinese are, for the most part, inclined to vegetarianism because of their religious beliefs
D. little land can be spared for pasture or forage crops

10. Which of the following is CHIEFLY responsible for the reduction in exports of Chilean nitrates? 10. A
A. Large-scale production of synthetic nitrates in the United States and Europe
B. The distance from the principal markets for nitrates
C. Depletion of the Chilean nitrate deposits
D. Discovery of competing nitrate deposits

11. Our GREATEST reserves of mineral fuels, as measured by their power to produce heat, consist of 11. D
A. coal
B. petroleum
C. natural gas
D. oil in shales

12. Which of the following comparisons of geographic conditions in the U.S. and Russia is NOT correct? 12. B
A. Russia is less richly endowed with ice-free coastline giving access to the oceans.
B. The chief rivers of Russia drain to the north, while those of the United States drain to the south.

C. More than half of the land area of Russia lies to the north of the most northern boundary of the U.S.
D. A relatively much larger proportion of the total land area of Russia is made up of mountain range and rough upland.

13. Prior to World War II, United States production MOST exceeded United States consumption of 13. B
A. iron ore B. bauxite C. zinc D. sulphur

14. The belt from about 20 degrees to 30 degrees latitude includes MAINLY 14. B
A. tropical jungles
B. subhumid and desert regions
C. coniferous forests
D. rainy grasslands

15. Which combination of objectives applies equally to the Missouri Valley Authority and to the Tennessee Valley Authority? 15. D
I. Flood control
II. Irrigation
III. Power development
IV. Recreation

The CORRECT answer is:
A. I, III, IV
B. I, II, IV
C. II, III, IV
D. I, II, III

16. In which of the following pairs is the direction of flow of the rivers approximately the SAME? 16. D
_____ River and _____ River.
A. Mississippi; Nile
B. Amazon; Danube
C. Volga; Rhine
D. Yangtse; Ohio

17. A study of the world's supply of coal would reveal that 17. A
I. there is a dearth of coal in the Southern Hemisphere
II. North America ranks first and Asia second in known reserves of coal
III. present known reserves of coal indicate the need for developing other sources of power within the next hundred years
IV. anthracite coal is fairly evenly distributed throughout the Northern Hemisphere

The CORRECT answer is:
A. I, II B. II, III C. I, IV D. III, IV

18. The steel industry is generally located near coal PRINCIPALLY because 18. C
A. most coal mines are in the vicinity of seaports
B. the iron needed in the manufacture of steel is usually found in deposits close to coal mines
C. the bulk of the coal needed to make steel is many times the bulk of iron
D. labor is cheaper in coal areas than in iron areas

19. The technique MOST frequently employed in dry farming is 19. B
 A. silting B. fallowing
 C. flood-watering D. irrigating

20. The diurnal temperature range is GREATEST in 20. A
 A. tropical deserts
 B. humid continental regions
 C. areas of Mediterranean climate
 D. tropical rainforests

21. Possession by an area of resources of salt, sulphur, lime, and coal makes possible industrial development in 21. B
 A. plastics B. chemicals
 C. glassware D. metal alloys

22. In connection with which of the following products has the rivalry between producers utilizing basically different climatic environments been most prolonged and significant? 22. D
 A. Coffee B. Beef cattle
 C. Natural rubber D. Sugar

23. Transatlantic planes flying from North America to Europe follow the _____ and, returning to America, follow the _____. 23. A
 A. Westerlies; Northeast Polar Winds
 B. Westerlies; Southeast Trade Winds
 C. Northeast Polar Winds; Westerlies
 D. Westerlies; Northwest Polar Winds

24. The fact that New York early surpassed Boston, Philadelphia, and Charleston in importance as a port is BASICALLY due to 24. B
 A. the superiority of New York harbor over other east coast ports
 B. its early ability to tap a larger hinterland than other east coast ports
 C. the concentration of shipping interests that developed from the fishing industry
 D. the network of railroads that developed around the city

25. Which of the following are characteristics of regions of Mediterranean climate? 25. D
 I. They are located in the middle latitude areas.
 II. They have drought in winter and rain in summer.
 III. Wheat is a typical crop
 IV. They employ summer irrigation where water is available.
 V. They have an absence of trees.

The CORRECT answer is:
 A. III, IV, V B. I, II, V
 C. I, II, III D. I, III, IV

KEY (CORRECT ANSWERS)

1. A
2. D
3. C
4. B
5. D
6. B
7. D
8. D
9. D
10. A
11. D
12. B
13. B
14. B
15. D
16. D
17. A
18. C
19. B
20. A
21. B
22. D
23. A
24. B
25. D

TEST 4

DIRECTIONS: Each question or incomplete statement is followed by several suggested answers or completions. Select the one that BEST answers the question or completes the statement. *PRINT THE LETTER OF THE CORRECT ANSWER IN THE SPACE AT THE RIGHT.*

1. Of the following, the MOST complete and accurate description of the pattern of land utilization on the Great Plains of the United States is: 1. C
   A. Wheat farming in the region east of the one hundredth meridan, commercial grazing west of that line, wheat forage crops and sugar beets in the irrigable districts
   B. Commercial grazing throughout the region except in years when unusual demand for wheat makes production of that crop profitable using dry farming methods
   C. Commercial grazing in the sections not suited for crops, wheat farming in the humid margins, forage crops and sugar beets in the irrigable districts
   D. Commercial grazing concentrated in the southern part of the region, wheat farming concentrated in the more humid northern part, forage crops and sugar beets in the irrigable districts

2. The Corn Belt of the United States 2. B
   I. is characterized by farms that average about 400 acres in size
   II. is bounded on the north by the hay and dairying belt
   III. produces hay, wheat, oats, soybeans, and tobacco as well as corn
   IV. is, in part, the prairie region reached by early settlers from the east
   V. is a region that employs crop rotation to maintain soil fertility

   The CORRECT answer is:
   A. I, II, III B. II, III, IV, V
   C. III, IV, V D. I, II, V

3. The spring wheat region of the United States includes parts of the states of 3. A
   A. Minnesota, North Dakota, South Dakota, Montana
   B. Texas, Oklahoma, Kansas, Nebraska
   C. West Virginia, Ohio, Indiana, Illinois
   D. Washington, Oregon, Idaho, Montana

4. Which of the following CORRECTLY associates physical features of the Intermontane area of the United States with related irrigated oases? 4. D

I. Colorado River - Imperial Valley
II. Wasatch Mts. - eastern Utah
III. Snake River - southern Idaho
IV. Salt River - central Arizona
V. San Joaquin River - Great California Valley

The CORRECT answer is:
A. I, III, IV, V
B. I, II, IV
C. II, III, V
D. I, II, III, IV

5. The development of manufacturing in the Piedmont region of the southern part of the United States is due PRIMARILY to the 5. B
A. presence there of cotton and tobacco for processing
B. abundant water power and cheap labor of the region
C. fact that the region is unsuited to typical southern crops, cotton and tobacco
D. accessibility of coal for power

6. The *Big Inch* pipeline 6. C
A. was abandoned after the close of World War II
B. carries petroleum from Texas and Oklahoma to New York and Philadelphia
C. carries natural gas from Texas to New York and Philadelphia
D. carries petroleum from Ohio and Indiana to Chicago

7. The Welland Canal connects Lake _____ and Lake _____. 7. A
A. Erie; Ontario
B. Superior; Michigan
C. Huron; Erie
D. Michigan; Huron

8. Which of the following statements concerning Alaska are TRUE? 8. D
I. The rugged topography of the Alaskan panhandle has made it a sparsely populated region despite the fact that its climate is milder than the other regions of Alaska.
II. The fishing industry employs more labor than the mining industry but the products of mining exceed those of fishing in value.
III. The food exports from the United States to Alaska are only those crops and products that cannot be produced in Alaska because of handicaps of soil and climate.
IV. The Matanuska Valley experiment proved the practicability of dairying and vegetable farming in Alaska.
V. Deposits of petroleum, anthracite coal, and bituminous coal are known to exist in Alaska.

The CORRECT answer is:
A. I, II, V
B. II, III, IV
C. III, V
D. IV, V

9. Recent investigations in Canada have revealed significant deposits of 9. C
   A. nickel
   B. copper
   C. iron ore
   D. zinc

10. Which of the following factors account for the comparative isolation of the lower St. Lawrence Valley? 10. C
    I. The river is ice bound for five to six months a year.
    II. Subsistence farming has increased in the last generation.
    III. There are few good harbors along the river.
    IV. Heavy snows make transportation difficult in winter.
    V. Deforestation has caused a rapid decline of the pulp and paper industry.

    The CORRECT answer is:
    A. I, II, III, V
    B. I, II, IV, V
    C. I, III, IV
    D. II, III, IV, V

11. Which one of the following statements is CORRECT concerning Arctic America? 11. B
    A. Some plant life grows but the environment permits production of nothing of economic value.
    B. The mid-summer days become quite warm or even hot.
    C. The lands are cold deserts.
    D. The reindeer is the only animal that exists in large numbers in this region.

12. Which one of the following has been characteristic of European economy during the past century? 12. B
    A. Its once moderately fertile soils have become exhausted.
    B. Soils unsuited for crops or pasture have been planted to forests.
    C. The loss through depletion of its mineral resources has been more than recouped by the discovery of new mines.
    D. Its sparsely distributed water power has been left largely unharnessed.

13. A person touring the industrial cities of Great Britain would find the wool industry centered at 13. A
    A. Leeds and Bradford
    B. Liverpool and Manchester
    C. Sheffield and Edinburgh
    D. Birmingham and Glasgow

14. Which one of the following statements concerning mineral resources in China is NOT true? 14. C
    A. Large deposits of coal are known to exist in China.
    B. China is relatively deficient in iron ore.
    C. Lack of tungsten and antimony resources is an important barrier to industrialization.
    D. Iron ore deposits are generally distant from coking coal deposits.

15. The region in Asia which supports the GREATEST density of agricultural population is 15.D
A. Malaya B. Ceylon C. China D. Java

16. Which of the following statements concerning Iran are CORRECT? 16.C
It
I. is a region of great aridity, much of it actual desert
II. has common frontiers on the west with Turkey, Syria, Iraq, and Saudi Arabia
III. has common frontiers with Russia, Afghanistan, and Pakistan
IV. borders on the Caspian Sea and the Persian Gulf

The CORRECT answer is:
A. I, II, III B. II, III, IV
C. I, III, IV D. I, II, IV

17. Which one of the following was the geographic factor MOST directly pertinent to the dispute over Kashmir? 17.C
A. Need for a land bridge between West Pakistan and East Pakistan
B. Absence of coal and mineral resources in India
C. Pakistan's need for water for irrigation
D. India's need for raw materials for her jute mills

18. Which of the following may CORRECTLY be considered part of the monsoon lands of Asia? 18.D
I. Indonesia II. Japan
III. Ceylon IV. Northeastern China
V. Southeastern China

The CORRECT answer is:
A. I, III, V B. I, II, V
C. III, IV D. I, II, III, IV

19. Which group consists ENTIRELY of European countries wherein the percentage of population engaged in agriculture is less than the percentage of the population of the United States that is engaged in agriculture? 19.C
A. France, Norway, Germany
B. England and Wales, Italy, the Czech Republic, and Slovakia
C. Belgium, Holland, Switzerland
D. Portugal, Sweden, Austria

20. The Campos of Brazil differ most markedly from the Llanos of Venezuela in 20.A
A. topography B. climate
C. natural vegetation D. land utilization

21. Which one of the following generalizations concerning the pattern of agriculture in South America is MOST accurate? 21. C
   A. Subsistence agriculture is characteristic of the lowlands near the coast, while the commercial plantations are in the tropical highlands.
   B. The vast expanse of the Andean highlands, generally not usable for agriculture, crowds plantations and subsistence farms into the coastal lowlands.
   C. The principal plantation areas are along the north and eastern coats, while the important subsistence agriculture regions are in the highland areas.
   D. With the exception of the coffee plantations in the highlands, most plantation agriculture and practically all subsistence agriculture are carried on in the lowland regions.

22. The agricultural products of the Central Valley of Chile MOST closely resemble those of 22. C
   A. southern Iran
   B. Vancouver Island
   C. the Cape district of Africa
   D. the coastal plain of Queensland

23. Which one of the following can be MOST correctly cited as a geographic reason for the delayed development of Africa? 23. A
The
   A. nature of the topography
   B. dense forests that cover the major part of the continent
   C. generally unfavorable climatic conditions throughout the continent
   D. general absence of fertile soil in regions otherwise suitable for agriculture

24. Which one of the following deserts owes its aridity to a combination of causal factors DIFFERENT from that of the other three? 24. D
_____ Desert.
   A. Sahara
   B. Atacama
   C. Australian
   D. Gobi

25. Which of the following regions of the world can PROPERLY be grouped as illustrating basically similar utilization of the natural environment? 25. A
   I. Northern Africa
   II. Patagonia
   III. Sinkiang
   IV. Arabia
   V. Mongolia

The CORRECT answer is:
   A. I, III, IV, V
   B. II, IV, V
   C. III, V
   D. I, III, IV

KEY (CORRECT ANSWERS)

1. C
2. B
3. A
4. D
5. B

6. C
7. A
8. D
9. C
10. C

11. B
12. B
13. A
14. C
15. D

16. C
17. C
18. D
19. C
20. A

21. C
22. C
23. A
24. D
25. A

TEST 5

DIRECTIONS: Each question or incomplete statement is followed by several suggested answers or completions. Select the one that BEST answers the question or completes the statement. *PRINT THE LETTER OF THE CORRECT ANSWER IN THE SPACE AT THE RIGHT.*

1. In the production of which of the following groups of strategic materials is the United States MOST deficient? 1. B
   A. Zinc, lead
   B. Chromium, manganese
   C. Copper, molybdenum
   D. Aluminum, iron ore

2. Which of the following basic materials is NOT correctly paired with the states that are important producers of the commodity? 2. B
   A. Crude petroleum - California, Texas
   B. Iron ore - Minnesota, Mississippi
   C. Steel - Ohio, Pennsylvania
   D. Wheat - North Dakota, Iowa

3. Which of the following statements are TRUE about shifts in the population of the United States? 3. B
   I. The states on the Pacific coast are the fastest growing states in the country.
   II. Since the 1940 census, the Middle Atlantic States have been losing population.
   III. California, with a population of about 49,000,000, is the most populated state in the United States.
   IV. In 1973 more than half of the American population was living in urban areas.
   V. In the last decade, our birth rate has been rising while our death rate has been falling.

   The CORRECT answer is:
   A. IV. V
   B. I, III, IV, V
   C. I, II, IV
   D. I, IV, V

4. Which combination presented, in the 1950's, the BEST case for the plan of a Columbia Valley Authority over plans proposed for the Missouri River and the St. Lawrence River Basin? 4. D
   The
   I. power shortage in the Northwest was the most acute in the nation
   II. Northwest had hitherto been treated as a *stepchild* in the appropriations of funds for conservation purposes
   III. CVA area was economically homogeneous
   IV. population of the CVA area was increasing at a rapid pace
   V. Columbia River Basin drained a larger area than did either the Missouri or the St. Lawrence

The CORRECT answer is:
A. I, II, V
B. I, II, IV
C. II, III, IV, V
D. I, II, III, IV, V

5. Which of the following statements about Canada are TRUE? 5. B
I. In manufacturing, Canada is surpassed today only by the United States, Great Britain, and Russia.
II. Canada's population is smaller than that of the United States but larger than that of any other nation in the western hemisphere except Brazil.
III. Canada surpasses continental United States in the following areas: wood exports, nickel, asbestos.
IV. Canada normally sells more to the United States than she buys from the United States.
V. Canada normally sells more to the United Kingdom than she buys from the United Kingdom.

The CORRECT answer is:
A. II, III, IV
B. I, III, V
C. II, III, V
D. I, II, V

6. Which of the following countries secures the HIGHEST yield of wheat per acre of land? 6. A
A. England
B. Argentina
C. Russia
D. United States

7. Which of the following comparisons of Great Britain with Japan are TRUE? 7. B
I. Both nations must today import most of their food supply.
II. Since World War II, both have been faced by the loss of important markets and sources of raw materials.
III. Japan's population density per square mile is from two to three times that of Great Britain.
IV. Both are large importers of raw cotton and manufacturers of cotton textiles.
V. Before World War II, both nations ranked about the same in the production of iron and steel.

The CORRECT answer is:
A. I, III, V
B. I, II, IV
C. II, IV, V
D. II, IV

8. The Irrawaddy delta, Rangoon, and the racial group of Karens are ALL associated with the country of 8. B
A. Sri Lanka B. Myanmar C. Korea D. India

9. Polar projection maps 9. A
A. distort areas at the equator
B. are used for navigation by transatlantic ocean liners
C. show latitude as straight lines
D. show the climate regions of the world

10. Which of the following generalizations concerning industrial locations is LEAST accurate? 10. B
   A. Smelters to process ores are usually located fairly near the ore producing mines.
   B. Refineries that handle crude metal are usually located near the ore producing mines.
   C. Petroleum is sometimes refined at the oil fields, sometimes near markets.
   D. Location of textile mills bears little relation to the location of the source of the fibers that they process.

11. In the production of pig iron in the blast furnace, the BASIC ingredients which are combined with the iron ore are 11. D
   A. coke and tungsten
   B. limestone and manganese
   C. tunsten and manganese
   D. coke and limestone

12. Which one of the following methods is NOT used to prevent erosion on hillsides? 12. D
   A. Erracing
   B. Contour plowing
   C. Trip cropping
   D. Gullying

13. Which of the following are characteristics of hydro-electric power? 13. A
   I. The source of supply is inexhaustible.
   II. Most rivers which are adapted for power are close to the markets for power.
   III. Transmission costs constitute a small fraction of the total cost of power.
   IV. The proportion of government owned plants is increasing.
   V. It is now the chief source of energy for American industry.

   The CORRECT answer is:
   A. I, IV
   B. I, II, III, IV
   C. II, III, IV, V
   D. I, IV, V

14. Nomadic herding is still a basic pattern of land utilization in 14. D
   I. northern Africa
   II. southwestern Asia
   III. Russian Turkestan
   IV. northern Scandinavia
   V. southeastern Australia

   The CORRECT answer is:
   A. I, II, V
   B. II, V
   C. I, III, IV
   D. I, II, III, IV

15. Which of the following ocean currents is located CORRECTLY? 15. D
   A. Labrador - Pacific Ocean side of North America
   B. Gulf - Atlantic Ocean side of South America
   C. Japanese - Indian Ocean side of Asia
   D. Humboldt - Pacific Ocean side of South America

16. Which one of the following deserts is INCORRECTLY associated with a country or continent? 16. C
   A. Kalahari - Africa B. Victoria - Australia
   C. Thar - Mongolia D. Atacama - Chile

17. Which of the following are characteristics of the tropics? 17. A
   I. The length of the days and nights is almost equal throughout the year.
   II. The sun's rays are never far from the perpendicular.
   III. The seasonal variations in temperature are not great.
   IV. There is practically no seasonal variation in rainfall.

   The CORRECT answer is:
   A. I, II, III B. II, III, IV
   C. I, III, IV D. I, II, III, IV

18. Which one of the following crops is associated with an area UNSUITED to its production? 18. A
   A. Corn - most parts of Canada
   B. Rye - sandy plains of Europe
   C. Barley - northern Europe
   D. Rice - Po Valley of Italy

19. A European city on approximately the same parallel of latitude as New York City is 19. B
   A. Oslo B. Madrid C. London D. Paris

20. The Continental Divide of the United States is 20. B
   A. a line down the mathematical center of the United States
   B. a line separating the river systems that flow to the east and west, respectively
   C. the Mississippi River
   D. a line separating North and South America

21. On which of the following Great Lakes does New York State border? 21. C
   Lake
   I. Ontario II. Erie III. Huron IV. Superior

   The CORRECT answer is:
   A. I, II, IV B. I, II, III
   C. I, II D. II, III, IV

22. In which of the following pairs is Canada the LEADING producer of the world? 22. C
   A. Petroleum and natural gas
   B. Wheat and corn
   C. Asbestos and nickel
   D. Wood pulp and rayon

23. The fact that fruit can be grown as far north in the United States as the shores of Lake Michigan is due to the 23. C
   A. development of artificial heating devices for protection from frost
   B. development of frost resisting varieties of fruits
   C. effect of the lake upon the climate of the adjoining district
   D. fact that the huge market for fruits in that region makes their cultivation profitable even after allowance for possible frost damage

24. Of the following agricultural belts, the one which lies FARTHEST north in the United States is the _____ belt. 24. C
   A. winter wheat
   B. cotton
   C. spring wheat
   D. corn

25. Which of the following statements concerning the Great Plains region of the United States are CORRECT? 25. D
   I. It is a region of grazing and agriculture.
   II. Rainfall is less than twenty inches annually throughout most of the area.
   III. It accounts for a large part of the beet sugar production of the nation.
   IV. The *Dust Bowl* is a part of this region.

The CORRECT answer is:
   A. I, IV
   B. II, III
   C. I, II, III
   D. I, II, III, IV

KEY (CORRECT ANSWERS)

1. B
2. B
3. B
4. D
5. B

6. A
7. B
8. B
9. A
10. B

11. D
12. D
13. A
14. D
15. D

16. C
17. A
18. A
19. B
20. B

21. C
22. C
23. C
24. C
25. D

EXAMINATION SECTION
TEST 1

DIRECTIONS: Each question or incomplete statement is followed by several suggested answers or completions. Select the one that BEST answers the question or completes the statement. *PRINT THE LETTER OF THE CORRECT ANSWER IN THE SPACE AT THE RIGHT.*

1. Comparison of Pakistan and India reveals that — 1. A
   - A. Pakistan has fewer mineral resources than India
   - B. the Deccan Plateau is part of Pakistan
   - C. Pakistan's population is predominantly Hindu
   - D. Pakistan is more developed industrially than India

2. Similarities between Iraq and Iran include — 2. A
   - I. dependence of agriculture on irrigation
   - II. presence of productive oil fields
   - III. large areas of desert land
   - IV. borders that touch Israel

   The CORRECT answer is:
   - A. I, II, III
   - B. III, IV
   - C. I, II
   - D. I, II, III, IV

3. The island of Bahrein is of significant geographical interest because of its — 3. B
   - A. strategic position
   - B. petroleum resources
   - C. famous Moslem shrines
   - D. productive date and coffee crops

4. Which of the following are the SIGNIFICANT reasons why Sao Paulo is the leading coffee region of Brazil? — 4. D
   - I. Coffee grows best in the rich loamy soil found in Sao Paulo.
   - II. The altitude is well suited for coffee production.
   - III. The wet and dry seasons of the uplands are favorable to coffee growing.
   - IV. Governmental and private capital have encouraged coffee production.

   The CORRECT answer is:
   - A. I, III
   - B. II, III, IV
   - C. I, II, IV
   - D. I, II, III, IV

5. In the humid Pampa region of Argentina, — 5. A
   - A. grazing is being gradually replaced by other types of land use
   - B. cultivation of wheat and corn is being gradually replaced by grazing
   - C. the hot moist climate has been utilized for coffee production
   - D. cotton is rapidly becoming the dominant crop

6. Development of an iron and steel industry in Brazil is hampered by the fact that 6. B
   A. Brazil lacks iron ore
   B. Brazil has iron ore but no coal
   C. the coal deposits of Brazil are far from the iron ore deposits
   D. Brazil lacks iron ore and coal

7. The United States and Russia differ MOST markedly in that 7. B
   A. Russia lacks important coal reserves; the United States has large coal reserves
   B. the United States has an excellent system of natural waterways; Russia has not
   C. Russia includes an extensive semi-arid region; the United States has no region of similar climate
   D. the United States is a cotton exporting country; Russia imports almost all the cotton it requires

8. Because of the distribution of natural resources in Europe, 8. C
   A. France must import bauxite
   B. Germany can export iron ore
   C. Sweden can export iron ore
   D. Denmark can export coal

9. One of the most important economic ties which link Turkey and the United States today is the fact that Turkey is a MAJOR supplier of 9. B
   A. bauxite B. chrome C. platinum D. titanium

10. Land utilization in Syria and Israel reflects the fact that 10. C
   A. the mountains are so close to the sea that there is no arable land along the coast
   B. the climate is too dry for cereal crops
   C. the mountains are a source of water for irrigation
   D. grazing is more profitable than agriculture in so dry a climate

11. The Saar is 11. A
   A. south of the Ruhr
   B. adjacent to the Ruhr
   C. a region rich in iron ore
   D. the source of the coal used in the Ruhr industries

12. In China 12. A
   A. rice is a dominant crop throughout the country
   B. in almost all parts of the country most of the rainfall occurs during the summer months
   C. the winters are generally mild
   D. about one-quarter of the land is not suitable for agriculture

13. The Sudan region of Africa is 13. C
   A. the important mineral producing region of the continent
   B. the rain forest region of the continent
   C. primarily a region of subsistence agriculture and grazing
   D. the principal plantation agriculture region of the continent

14. Homolographic map projections are characterized by 14. D
   A. distortion of area but accuracy of shape
   B. accurate representation of area
   C. centering of the map around a polar area
   D. accurate representation of shape and area

15. Of the following pairs, the one which does NOT represent regions of similar climate is 15. D
   A. southeastern United States - southern China
   B. central Europe - northeastern United States
   C. southern Chile - western Europe
   D. central Australia - New Zealand

16. Most of the world's people live in lands with an annual rainfall of _____ inches. 16. B
   A. 0-20 B. 20-80 C. 80-120 D. over 120

17. Winter wheat is planted in 17. D
   A. early summer, harvested in late autumn
   B. late summer, harvested just before frost
   C. spring, harvested in autumn
   D. autumn, harvested in early summer

18. Of the following, the one NOT a hardwood is 18. C
   A. oak B. elm C. pine D. walnut

19. Of the following areas, the one NOT in approximately the same range of latitude as the others is 19. D
   A. Newfoundland B. the Caspian steppes
   C. northern France D. the Scottish Highlands

20. Millet is an important product of 20. B
   A. the area north of the Sahara desert
   B. northern China
   C. southern China
   D. the Iranian plateau

21. Of the following, the one NOT a leading manufacturing region of Europe is 21. B
   A. the Midlands B. Brittany
   C. the Saar D. the Po Plain

22. Of the following, the one which has NOT favored the development of western Europe's dairy industry is the 22. A
    A. growth of urban population in the region
    B. cool and moist summers in the region
    C. availability of cheap imported feed for livestock
    D. lack of suitable land for forage crops

23. The tundra belt of Russia is characterized by 23. A
    A. mosses, grasses, and stunted shrubs
    B. evergreen forests of considerable commercial value
    C. level, gently rolling surface suitable for farming
    D. desert land suitable for irrigation projects

24. The famous fishing bank known as the Dogger Bank is located 24. D
    A. off Newfoundland B. in the Sea of Japan
    C. off Cape Cod D. in the North Sea

25. Of the following, the one who did NOT make an oustanding contribution to the study of the influence of geography on human affairs is 25. A
    A. Hugo Munsterberg B. Ellen G. Semple
    C. Ellsworth Huntington D. Isaiah Bowman

KEY (CORRECT ANSWERS)

1. A	11. A
2. A	12. A
3. B	13. C
4. D	14. D
5. A	15. D
6. B	16. B
7. B	17. D
8. C	18. C
9. B	19. D
10. C	20. B

21. B
22. A
23. A
24. D
25. A

TEST 2

DIRECTIONS: Each question or incomplete statement is followed by several suggested answers or completions. Select the one that BEST answers the question or completes the statement. *PRINT THE LETTER OF THE CORRECT ANSWER IN THE SPACE AT THE RIGHT.*

1. Of the following cities, the one NOT in the hinterland of the ports of Amsterdam and Rotterdam is 1. D
   A. Frankfort B. Basle C. Munich D. Danzig

2. Of the following, the ones which are NOT causal factors in the economic problems of Italy are 2. A
   I. climate unsuited to the production of wheat
   II. lack of coal
   III. inadequate agricultural land
   IV. lack of water power sites

   The CORRECT answer is:
   A. I, IV B. II, III
   C. II, III, IV D. I, III

3. Of the following, the one NOT a natural geographic region of Argentina is 3. C
   A. the Chaco B. the Pampa
   C. Patagonia D. the Atacama Desert

4. Of the following, the one which does NOT account for the large agricultural production of India is the 4. D
   A. large area of arable land
   B. dependable rainfall of the winter monsoon
   C. dense agricultural population
   D. extensive use of irrigation

5. Which two of the following regions of Canada are MOST heavily populated? 5. C
   The
   I. Maritime Provinces II. Great Interior Plain
   III. Laurentian Shield IV. St. Lawrence Valley

   The CORRECT answer is:
   A. I, II B. II, III C. I, IV D. II, IV

6. The Sudan in Africa is an example of the climate type known as 6. B
   A. tropical rainforest B. savanna
   C. humid sub-tropical D. desert

7. Despite heavy emigration from Europe to other parts of the world, Africa has remained relatively sparsely populated because of the 7. B

I. prevalence of the tsetse fly
II. barrier of the desert and escarpment
III. difficulty of building roads or railroads from the coast into the interior
IV. hot and humid climate throughout most of the continent

The CORRECT answer is:
A. I, II, III B. II, III, IV
C. I, III, IV D. I, II, IV

8. Of the following major geographic regions of the United States, the one which forms a transition zone between AT LEAST two of the other three regions is the 8. A
A. Great Plains B. South
C. Intermountain West D. Middle West

9. Before World War II, the principal countries from which the United States imported tin included all of the following EXCEPT 9. C
A. British Malaya B. Netherlands East Indies
C. China D. Bolivia

10. Truck farming has expanded rapidly in the South owing to 10. A
I. cheap labor
II. boll weevil destruction of cotton
III. increase in southern demand for truck farm products
IV. use of river transportation facilities
V. climate which permits production when northern farms cannot supply their markets

The CORRECT answer is:
A. I, III, V B. I, II, V
C. II, III, IV D. I, II, IV

11. Of the following, the one NOT a factor in Pittsburgh's industrial leadership is 11. D
A. the Connellsville bed of coking coal
B. Pittsburgh's proximity to Great Lakes ports
C. the rich veins of iron ore in Pittsburgh's immediate vicinity
D. Pittsburgh's location in relation to the Corn Belt

12. The CHIEF money crop produced in the United States Corn Belt is 12. B
A. winter wheat
B. corn for livestock
C. corn for human consumption
D. dairy products

13. Of the following regions, which is the center for the production of a type of coal not produced in the other three regions? 13. B
A. Iowa B. Eastern Pennsylvania
C. Illinois D. Eastern Ohio

14. Which one of the following statements about Indo-China is INCORRECT? 14. A
    A. Indo-China is made up of the states of Laos, Cambodia, and Viet Minh.
    B. Its mountain chains are heavily forested and extensive.
    C. The Mekong and Red Rivers are major streams.
    D. Dienbienphu is near the border of Laos.

15. With the exception of Indonesia, the nations of the western Pacific are economically and militarily handicapped by a lack of 15. C
    A. coal
    B. tin
    C. petroleum
    D. water power

16. United States *economic* interest in Southeast Asia may BEST be accounted for by the fact that this area mines a significant amount of the world's 16. A
    A. tungsten and tin
    B. coal and iron ore
    C. gold and silver
    D. lead and zinc

17. The climate of Japan is similar to that of the _____ coast of _____. 17. A
    A. eastern; the United States
    B. eastern; Canada
    C. eastern; Mexico
    D. western; Canada

18. The littoral area of the North Atlantic Coastal Region derives its economic importance from the fact that 18. D
    A. the New England farmer is making extensive use of machinery in growing corn
    B. Maine containues to be of outstanding importance in the production of lumber
    C. gypsum, salt, and limestone mining are key industries in the New England states
    D. recreation as a business is playing an increasingly predominating role

19. China has a common boundary with 19. B
    A. India, Korea, Malaya, Thailand
    B. Burma, Vietnam, Korea, Russia
    C. India, Iran, Korea, Thailand
    D. Afghanistan, Iran, Korea, Russia

20. Which one of the following rebellions stemmed MOST directly from the influences of *geographic* factors on history? 20. C
    _____ Rebellion.
    A. Shays'
    B. Nat Turner's
    C. The Whiskey
    D. Bacon's

21. The growth of the aluminum industry in the Pacific North-west can MOST readily be attributed to 21. B
   A. nearness to market
   B. development of hydro-electric power
   C. increased labor supply
   D. suitable climate

22. The fact that Seattle, Washington, has a growing season of 246 days, while Atlanta, Georgia, has a growing season of 225 days can BEST be explained by the 22. C
   A. latitude of Seattle as compared with the latitude of Atlanta
   B. longitude of Seattle as compared with the longitude of Atlanta
   C. marine climate of Seattle
   D. humid subtropical climate of Atlanta

23. A city which is NOT on the St. Lawrence Seaway is 23. C
   A. Ogdensburg B. Massena
   C. Rochester D. Buffalo

24. The city which does NOT produce any significant quantity of steel is 24. D
   A. Gary B. Buffalo C. Detroit D. Scranton

25. Which of the following statements regarding Pakistan BEST describe the difference between the economy of Pakistan and that of Bangladesh? 25. C
   I. The staple food of Bangladesh is rice.
   II. Bangladesh has the greater population.
   III. Pakistan's chief export is jute.
   IV. Bangladesh has recurrent droughts.
   V. Pakistan's area is the greater.

   The CORRECT answer is:
   A. I, II B. I, III, IV
   C. I, II, V D. II, IV, V

KEY (CORRECT ANSWERS)

1. D	6. B	11. B	16. A	21. B
2. A	7. B	12. B	17. A	22. C
3. C	8. A	13. B	18. D	23. C
4. D	9. C	14. A	19. B	24. D
5. C	10. A	15. C	20. C	25. C

TEST 3

DIRECTIONS: Each question or incomplete statement is followed by several suggested answers or completions. Select the one that BEST answers the question or completes the statement. *PRINT THE LETTER OF THE CORRECT ANSWER IN THE SPACE AT THE RIGHT.*

1. Which one of the following groups is INCORRECTLY paired? 1. C
   A. Lake Superior and Lake Huron - Sault Sainte Marie Canal
   B. Lake Erie and Lake Ontario - Welland Canal
   C. Bay of Biscay and North Sea - Kiel Canal
   D. Mediterranean Sea and Red Sea - Suez Canal

2. The term *heartland* in writings on geopolitics refers to the 2. A
   A. center of the Eurasian land mass
   B. western hemisphere
   C. World Island
   D. North Atlantic Treaty Nations

3. After World War II, Italy's power output was rebuilt by 3. A
   A. Italy's neglected coal resources
   B. the Alpine water power potential
   C. Italy's petroleum potential
   D. the shift of the tides in the Bay of Naples

4. Which one of the following seas does NOT border on Russia? 4. D
   A. Sea of Japan
   B. Baltic
   C. Black
   D. Mediterranean

5. South America lags in manufacturing and heavy industries because it 5. A
   A. lacks large population near high-grade coal and iron reserves
   B. is farther away from Europe than from North America
   C. lacks oil reserves near available transport
   D. lacks potential hydro-electric power

6. Which one of the following does NOT describe a great circle? 6. D
   A. A plane passing through the center of a sphere
   B. The meridians
   C. The shortest distance between two points on the surface of a sphere
   D. Parallels of latitude

7. All of the following are characteristics of the Arctic tundra EXCEPT 7. B
   A. wide annual extremes of temperature
   B. constant high winds
   C. slight daily variations of temperature
   D. less annual precipitation than in many desert areas

8. On Mercator's projection, a straight line would MOST nearly represent the shortest distance between New York City and 8. D
   A. Rome B. Moscow
   C. Tokyo D. Santiago, Chile

9. Coal, steel, chemicals, and textiles are MOST heavily concentrated in the 9. C
   A. Saar Valley B. Lorraine Province
   C. Ruhr district D. Donetz Basin

10. Which one of the following statements concerning the various kinds of rubber is CORRECT? 10. D
   A. Wild rubber is gradually regaining its former predominant position in the industry.
   B. Synthetic rubber has dropped to the relative position it had prior to World War II.
   C. Natural rubber is cheaper than and superior in quality to synthetic rubber.
   D. Plantation rubber has replaced wild rubber because of savings in labor and transportation.

11. The St. Lawrence Seaway 11. B
   A. decreases the importance of the *Soo* Canals
   B. enables large vessels to reach Chicago, Detroit, and Cleveland from the Atlantic Ocean
   C. provides a toll-free passage to the midwest
   D. provides year-round transportation through the Great Lakes

12. In which of the following respects does the corn and winter wheat belt of the United States contrast MOST sharply with that of the Soviet satellites of southeastern Europe? 12. C
   I. Size of farms II. Type of crops
   III. Climate IV. Extent of machinery

   The CORRECT answer is:
   A. I, III, IV B. II, III
   C. I, IV D. I, II, IV

13. Which one of the following factors has NOT contributed to the pre-eminence of the Lake Superior iron mining district? 13. B
   A. Advantages of transportation
   B. Presence of coal in the same area
   C. Adaptability to the open pit method of mining
   D. Abundance of high grade ores

14. Which of the following factors have provided the impetus for the movement of industries from New England to the South? 14. A
   I. Tax concessions II. Labor conditions
   III. High tariffs IV. Freight rates

The CORRECT answer is:
A. I, II
B. I, II, III
C. II, IV
D. III, IV

15. Proximity to urban markets and existence of extensive hay production account for the location of which of the following industries? 15. C
A. Poultry raising
B. Hog production
C. Dairying
D. Truck farming

16. Which one of the following nations has the SMALLEST proportion of its population engaged in farming? 16. B
A. United States
B. Great Britain
C. France
D. Canada

17. Which one of the following statements regarding the principal humid subtropical areas is CORRECT? 17. B
A. They are found on the southwestern coasts of the continents.
B. They are found in approximately the same latitudes as the Mediterranean climate regions.
C. The greatest extent of this type of climate is found in Africa.
D. They experience a drought period during the winter.

18. The fact that subtropical crops are grown in the gulf coast states while temperate zone crops are grown farther south in Mexico is explained by differences in 18. C
A. latitude
B. longitude
C. altitude
D. irrigation

19. In which of the following respects are the Mississippi River and the Mackenzie River similar? 19. D
They
I. are nearly the same length
II. flow through wide lowlands
III. are suited to intensive farming
IV. have well-developed deltas

The CORRECT answer is:
A. I, II
B. II, III, IV
C. I, III
D. I, II, IV

20. Which one of the following factors has NOT contributed to the relative decrease of coal production in the United States? 20. A
A. Depletion of coal reserves
B. Interruptions caused by strikes
C. Increasing use of competing sources of power
D. More efficient use of coal

21. The lumber industry of the Pacific coast of the United States 21. D
I. has yielded to the South as the leading lumber producing area of the United States
II. is dependent upon the marine type of climate
III. has been in operation for over a century and a half
IV. produces timber from fir, spruce, and cedar trees

The CORRECT answer is:
A. I, II B. I, IV C. II, III D. II, IV

22. Two cities which owe their importance in LARGE measure to control of narrow channels or straits are 22. B
A. Chicago and New Orleans
B. Copenhagen and Istanbul
C. Panama and Caracas
D. Los Angeles and San Francisco

23. Which one of the following groups is INCORRECTLY paired? 23. C
A. Brenner Pass - Italy and Austria
B. The Iron Gate - Yugoslavia and Rumania
C. St. Bernard Pass - Switzerland and France
D. Khyber Pass - Pakistan and Afghanistan

24. The concentration of Australia's population in the south-eastern coastal region is explained by the prevalence of 24. B
A. a variety of mineral resources
B. a moist climate
C. large scale irrigation projects
D. suitable conditions for sheep raising

25. Which of the following are geographic handicaps that have retarded the development of Asiatic peoples? 25. D
I. Lack of first-class harbors
II. Absence of favorable routes leading to the interior
III. Desert climate experienced by a large part of the continent
IV. Vast extent of high plateaus

The CORRECT answer is:
A. I, II B. III, IV
C. I, IV D. I, II, III, IV

KEY (CORRECT ANSWERS)

1. C	6. D	11. B	16. B	21. D
2. A	7. B	12. C	17. B	22. B
3. A	8. D	13. B	18. C	23. C
4. D	9. C	14. A	19. D	24. B
5. A	10. D	15. C	20. A	25. D

TEST 4

DIRECTIONS: Each question or incomplete statement is followed by several suggested answers or completions. Select the one that BEST answers the question or completes the statement. *PRINT THE LETTER OF THE CORRECT ANSWER IN THE SPACE AT THE RIGHT.*

1. The significance of the 100th meridian in American agriculture is that it — 1. A
   A. closely parallels the 20-inch annual precipitation line
   B. divides the Corn Belt in half
   C. separates the subtropical crops from the temperate crops
   D. is the line beyond which only dry farming is practiced

2. Wheat farming in the United States is extensively mechanized because — 2. B
   A. the farmer is thereby assured a steadier income than is a farmer using mostly hand labor
   B. it is adapted to extensive farming featured by relatively cheap land and expensive labor
   C. wheat gives high yield per acre but low yield per man
   D. machine methods are best adapted to intensive farming

3. Of the following, the principle of vertical integration of industry is BEST illustrated by the — 3. A
   A. Ford plant at River Rouge
   B. New York City garment industry
   C. Gary plant of the Chicago-Illinois Steel Company
   D. General Foods Company

4. In traveling northward from the Gulf of Mexico to the state of Wisconsin, what is the CORRECT order in which one would pass through the following farming regions? — 4. A
   I. Humid subtropical crop belt
   II. Hay and dairy region
   III. Corn and winter wheat belt
   IV. Corn belt
   V. Cotton belt

   The CORRECT answer is:
   A. I, V, III, IV, II
   B. V, I, IV, III, II
   C. V, I, II, III, IV
   D. I, V, IV, III, II

5. In which of the following pairs has the first region *decreased* and the second region *increased* in percentage of American lumber production since 1870? — 5. C
   A. Southern states - Lake states
   B. Pacific states - Southern states
   C. Northeastern states - Pacific states
   D. Rocky Mountain states - Central and Prairie states

6. The strategic location of Labrador, Alaska, and Eastern Siberia in relation to American overseas air travel can BEST be shown on which of the following map projections? 6.B
   A. Mercator's
   B. North Polar Azimuthal Equidistant
   C. Goode's Interrupted Homolosine
   D. Orthographic

7. The dry subtropical climate is found in 7.B
   I. Mediterranean area
   II. southern California
   III. central Chile
   IV. Pacific northwest

   The CORRECT answer is:
   A. I *only*
   B. I, II, III
   C. I, III
   D. II, III, IV

8. If the British Foreign Minister calls the British Ambassador to the United Nations at noon, on Monday, London time, the time in Washington, D.C. is 8.C
   A. Monday Noon
   B. 4 P.M. Monday
   C. 7 A.M. Monday
   D. 6 P.M. Sunday

9. In which of the following foodstuffs is there competition between the temperate and tropic regions? 9.A
   A. Sugar
   B. Rice
   C. Wheat
   D. Corn

10. The MOST favorable natural conditions for the development of commercial fisheries are found in 10.B
    A. warm oceans
    B. zones of contact between warm and cold currents
    C. deep coastal waters
    D. cold ocean currents

11. The term *soil mining* refers to 11.A
    A. sale of the soil minerals in the form of agricultural products
    B. mining of iron ore by scooping it up from the surface of the land
    C. extraction of coal from a vein lying between two layers of soil
    D. extraction of petroleum from shale

12. The fact that nine-tenths of the world's rye is grown on the European plains from Belgium to Russia is BEST explained by the 12.B
    A. European taste for rye breads
    B. inferior soils of these plains
    C. extensive use of rye straw as a cover crop
    D. semi-arid climate of these plains

13. Subtropical vegetation found on the Italian Riviera may be explained by which of the following factors? 13.C
    I. Strong insolation received by the south-facing slopes
    II. Location close to the Tropic of Cancer
    III. Nearness to the Mediterranean
    IV. Protection from cold winds provided by the Alps

The CORRECT answer is:
A. I, II
B. II, IV
C. I, III, IV
D. II, III, IV

14. European manufacturing districts which would be harmed MOST by prolonged droughts are 14. D
I. Ruhr Valley
II. Silesia
III. Switzerland
IV. Po Valley

The CORRECT answer is:
A. IV *only*
B. II, III, IV
C. I, III
D. III, IV

15. A former satellite nation which BEST supplemented the petroleum production of the U.S.S.R. was 15. D
A. Hungary
B. Poland
C. Czechoslovakia
D. Rumania

16. In which of the following countries does the central region roughly resemble California in environment and products? 16. C
A. Peru B. Paraguay C. Chile D. Argentina

17. A geographic region in South America noted for its mineral wealth is the 17. C
A. tropical savanna
B. Pampas
C. Atacama Desert
D. Chaco

18. Desert oases play an important role in the economy of 18. B
A. Bolivia B. Peru C. Ecuador D. Colombia

19. Which statements CORRECTLY explain why it was possible to develop the Caribbean coast from Yucatan to Colombia into the so-called *Banana Coast*? 19. D
I. There are fertile delta soils in the region.
II. The rainfall in the region is abundant and occurs during winter and summer months.
III. The region is characterized by the alternation of wet winters and dry summers necessary for banana cultivation.
IV. The leached soils of the region respond readily to fertilization.

The CORRECT answer is:
A. III, IV B. II, IV C. I, III D. I, II

20. The influence of the mountainous topography of Japan is seen in 20. D
I. its climate
II. the Japanese diet
III. the type of power resources available
IV. the emphasis on timber resources

The CORRECT answer is:
A. I, III
B. III, IV
C. I, IV
D. I, II, III, IV

21. Attempts by United States entrepreneurs to introduce rubber production in Africa resulted in the creation of the world's largest single rubber plantation in 21.A
    A. Liberia B. Nigeria
    C. Gold Coast D. Ethiopia

22. The river which furnishes commerce and transportation to the GREATEST number of people is the 22.B
    A. Volga B. Yangtze C. Nile D. Rhine

23. Abundant harbors and unsuitability to agriculture combine to account for the fact that the world's third largest merchant fleet is owned by 23.A
    A. Norway B. France
    C. Netherlands D. Denmark

24. The seaport which owes much of its importance to the fact that it is located at the mouth of the Rhine is 24.B
    A. Amsterdam B. Rotterdam
    C. Bremen D. Hamburg

25. Significant characteristics of Australia's population include the facts that 25.D
    I. almost half the population lives in the cities
    II. the policy of unrestricted immigration has contributed materially to the rapid growth of the population
    III. practically all the large cities are within 100 miles of the coast
    IV. the population is approximately equal to that of New York City
    A. I,III B. II,III,IV C. I,II,IV
    D. I,III,IV

KEY (CORRECT ANSWERS)

1. A
2. B
3. A
4. A
5. C

6. B
7. B
8. C
9. A
10. B

11. A
12. B
13. C
14. D
15. D

16. C
17. C
18. B
19. D
20. D

21. A
22. B
23. A
24. B
25. D

TEST 5

DIRECTIONS: Each question or incomplete statement is followed by several suggested answers or completions. Select the one that BEST answers the question or completes the statement. *PRINT THE LETTER OF THE CORRECT ANSWER IN THE SPACE AT THE RIGHT.*

1. One advantage of the Mercator projection is that 1. B
   A. there is a minimum of distortion in shape of land masses as distance from the equator increases
   B. all parallels of latitude cross all meridians at right angles, as on a globe
   C. there is no distortion of land areas except north of the Arctic Circle
   D. the rhumb lines are curved lines

2. In which one of the following types of climate are the four seasons of the year MOST clearly identifiable? 2. A
   A. Humid Continental B. Humid Subtropical
   C. Mediterranean D. West Coast Marine

3. The value of the tropical wet regions of the world as areas for future settlement is a highly controversial question. Which one of the following is a *valid* and *pertinent* statement on this issue? 3. C
   A. All the tropical wet regions of the world are now densely populated.
   B. The soil in these regions is so rich that they have great potential value as sources of food.
   C. Soil conditions in these regions make the growth of crops, other than bush and tree crops, difficult.
   D. While daytime temperatures permit year-round agriculture, the temperature decline at night in these regions is so sharp as to jeopardize many food crops.

4. Most of the world's population is concentrated in areas that have an average annual rainfall ranging from 4. D
   A. 1 to 10 inches B. 11 to 20 inches
   C. 21 to 80 inches D. 81 inches and above

5. New York City lies on approximately the same parallel of latitude as 5. C
   A. Tokyo B. Moscow C. Rome D. Paris

6. Which one of the following choices is CORRECTLY paired? 6. B
   A. Pampas - semi-arid regions of middle latitudes
   B. Tundras - subarctic regions
   C. Tropical savannas - tropical rainforest throughout the whole year
   D. Steppes - sharply-rising, broken terrain of the low latitudes

7. The St. Lawrence Seaway 7. D
   A. has moved the transfer point from ocean-going to canal-going boats much farther inland
   B. controls the current of the St. Lawrence River as it flows toward the Great Lakes
   C. by-passes the old Welland Canal between Lake Ontario and Lake Erie
   D. makes accessible to ocean liners a coastline longer than the United States Atlantic seaboard

8. Of the following, the MOST important factor that has helped the United States become a first ranking nation in the production of iron and steel is 8. C
   A. a supply of cheap labor near the iron ore deposits
   B. the advantage of an early start
   C. the proximity of iron deposits to cheap water transportation
   D. a suitable climate in areas of large iron ore deposits

9. The TRUE core of the Dairy Belt in the United States is 9. A
   A. Wisconsin and Minnesota B. Michigan and Nebraska
   C. Iowa and Kansas D. Illinois and Indiana

10. The United States id dependent on foreign sources for its supply of 10. A
    A. chromium B. copper C. lead D. zinc

11. All of the following factors account for the prominence of Gary, Indiana, as a steel center EXCEPT 11. B
    A. accessibility to coal, iron, and limestone
    B. an underlying geological structure that provides an unusually strong foundation to support heavy steel plants
    C. nearness to a tremendous steel market, actual and potential
    D. a crossroads position relative to water and rail transportation

12. In the United States, the time zone lying between the 75th and 90th meridian is the _____ Time Belt. 12. A
    A. Eastern B. Pacific C. Mountain D. Central

13. Both the United States and Canada are DEFICIENT in 13. A
    A. manganese and tin
    B. nickel and asbestos
    C. livestock and natural gas
    D. bauxite and phosphates

14. Which one of the following accounts for the economic importance of Curacao and Aruba? 14. B
    The
    A. mining of uranium
    B. refining of oil from the Maracaibo oil fields of Venezuela
    C. growing of bananas, coffee, and rubber
    D. logging of rare and valuable tropical woods

15. In which pair are the economies of both countries LARGELY dominated by the Andean highlands of South America? 15. C
    A. Brazil and Venezuela B. Argentina and Paraguay
    C. Colombia and Peru D. Bolivia and Uruguay

16. Some modern geographers no longer conceive of Europe as a continent because 16. C
    A. modern methods of transportation and communication have eliminated barriers
    B. Europe and the Western Hemisphere are a cultural entity
    C. geographically, Europe is only a peninsula of Asia
    D. the discovery of the Arctic and of Antarctica have proven that there are no continents at all

17. Europe is the LEADING continent in the production of 17. C
    A. lead B. lumber C. coal D. petroleum

18. All of the following are industrial areas based on rich coal fields EXCEPT the 18. A
    A. Black Forest
    B. Donets Basin
    C. Midlands around Manchester and Birmingham
    D. Ruhr Valley

19. Before reunification, a comparison of the economics of West Germany and East Germany showed that East Germany had 19. C
    A. the greater population
    B. more heavy industry
    C. greater agricultural acreage
    D. no coal deposits

20. Which of the following countries are MOST similar in the commercial advantages of their relative locations? 20. A
    A. Great Britain and Japan
    B. United States and Germany
    C. Russia and Great Britain
    D. China and Japan

21. Which of the following pairs contains regions that are MOST different from each other in climate? 21. C
    A. Southern California - Spain
    B. Arizona - Central Australia
    C. Coast of Louisiana - Amazon Valley
    D. North Pacific coast of the United States - Western Europe

22. According to recent estimates, the LARGEST oil reserves in the world are located in 22. B
    A. the United States B. the Middle East
    C. South America D. Russia

23. Okinawa is the LARGEST island of the 23. D
    A. Caroline Islands
    B. Dutch East Indies
    C. Marshall Islands
    D. Ryukyu Chain of Islands

24. The efficiency of the plantation system is demonstrated by the successful transfer from South America to Indonesia of the major part of the production of 24. C
    I. cinchona
    II. rice
    III. rubber
    IV. quebracho

    The CORRECT answer is:
    A. I, IV
    B. III, IV
    C. I, III
    D. I, II, IV

25. In which of the following ways do China and India resemble each other economically? 25. D
    The
    I. minor importance of the domestic animal to the economy
    II. degree of development of rail transportation
    III. importance of river valleys to the economy
    IV. degree of development of heavy industry

    The CORRECT answer is:
    A. I, II
    B. II, III, IV
    C. I, III, IV
    D. I, III

KEY (CORRECT ANSWERS)

1. B
2. A
3. C
4. D
5. C

6. B
7. D
8. C
9. A
10. A

11. B
12. A
13. A
14. B
15. C

16. C
17. C
18. A
19. C
20. A

21. C
22. B
23. D
24. C
25. D

EXAMINATION SECTION

TEST 1

DIRECTIONS: Each question or incomplete statement is followed by several suggested answers or completions. Select the one that BEST answers the question or completes the statement. *PRINT THE LETTER OF THE CORRECT ANSWER IN THE SPACE AT THE RIGHT.*

1. The industry of France is handicapped by domestic deficiencies in 1. A
   A. petroleum
   B. iron ore
   C. water power
   D. productive soil

2. Which one of the following is CORRECT with regard to the Saar Valley? 2. C
   It
   A. is an internationalized area under the control of the European Iron and Steel Community
   B. is committed to exporting its entire coal production to France and Italy
   C. has become economically integrated with Germany, but ships coal to France
   D. is politically integrated with Germany but economically integrated with France

3. Before reunification, 3. B
   A. the borders of both East Germany and West Germany provided natural protection from armed invasion
   B. although East Germany was virtually self-sufficient in food production, West Germany imported one-third of its food requirements
   C. industrial production was approximately equally divided between West Germany and East Germany
   D. the population of both East Germany and West Germany was less than three-quarters of the pre-World War II population of Germany

4. Which one of the following is an example of a small nation that bases its livelihood almost entirely upon one industry? 4. C
   A. Belgium
   B. Ireland
   C. Luxembourg
   D. Netherlands

5. The greater portion of Russia's mineral resources is located in 5. C
   A. the area near the Arctic Circle
   B. Siberia
   C. Southern Russia and the Caucasus
   D. the Ural Mountains

6. Of the total area of the former U.S.S.R, 6. B
   A. practically all could have been used as farm land
   B. only about 25% could have been considered farm land
   C. most of the tillable soil was in Soviet Asia
   D. the acreage in crops was four times that of the United States

7. Monsoon Asia embraces 7. C
   A. Turkey, Arabia, Iraq
   B. Iran and Afghanistan
   C. India, Eastern China, Japan
   D. Western China and Tibet

8. Thailand is located between 8. A
   A. Myanmar and Cambodia
   B. India and China
   C. China and Malaya
   D. India and Malaya

9. Which one of the following is a CORRECT generalization concerning the Philippines? 9. B
   A. Their low cost agricultural products have brought them prosperity since their ties to American markets were released.
   B. Although rice is the staple food crop, it is an important export.
   C. Among the important crops are cacao and sisal.
   D. Japanese interest in the Philippines centered on the islands' important reserves of coal and petroleum.

10. The Hawaiian Islands are in the same latitude as 10. B
    A. Florida
    B. Cuba
    C. Panama Canal Zone
    D. California

11. Which one of the following regions of Africa is a significant producer of non-ferrous, non-precious metals? 11. B
    A. Sudan
    B. Katanga-Rhodesia
    C. Guinea Coast
    D. Madagascar

12. The African continent produces more than half of the world's supply of both 12. D
    A. copper and manganese
    B. wool and cotton
    C. coffee and natural rubber
    D. cacao and diamonds

13. An area whose lowlands are drained mostly by rivers running into the Caribbean is 13. C
    A. Paraguay, Uruguay, Argentina
    B. Ecuador, Peru, Chile
    C. Colombia, Venezuela, the Guianas
    D. Brazil

14. Which one of the following has hindered the industrial expansion of Argentina? 14. C
    A. Tropical climate which hinders production
    B. Lack of suitable seaports

C. Lack of basic resources such as coal and iron ore
D. Lack of railroad facilities within Argentina

15. Canada is a major producer of 15. A
A. nickel and cobalt B. tin and vanadium
C. petroleum and iron ore D. bauxite and asbestos

16. Which one of the following Canadian cities owes its rapid growth to its proximity to the Alcan Highway, to uranium deposits, and to its location with respect to recent oil discoveries? 16. B
A. Sudbury B. Edmonton C. Winnipeg D. Ottawa

17. Which one of the following cities in the United States is INCORRECTLY linked with one of its leading products? 17. B
A. Akron - tires B. Troy - shoes
C. Kansas City - flour D. Pittsburgh - steel

18. A ship entering the St. Lawrence Seaway at Montreal, en route to Chicago, would pass through Lakes 18. C
A. Erie, Ontario, Michigan, Huron
B. Huron, Erie, Ontario, Michigan
C. Ontario, Erie, Huron, Michigan
D. Erie, Huron, Michigan, Ontario

19. The economic prosperity of Arizona and of Utah is linked significantly to the demand for 19. B
A. gold B. copper C. silver D. petroleum

20. The two non-metallic minerals of which New York State has been a leading producer are 20. A
A. gypsum and salt B. salt and sulphur
C. coal and graphite D. phosphate and nitrate

21. Which one of the following statements about the doldrums is CORRECT? 21. A
A. The primary movement of air is upward.
B. The primary movement of air is downward.
C. The primary movement of air is easterly.
D. There is practically no movement of air.

22. Which one of the following terms is NOT used to describe types of vegetation? 22. A
A. Chernozem B. Steppe C. Chaparral D. Savanna

23. Which one of the following statements is TRUE? The length of one degree of longitude 23. C
A. never approximates that of one degree of latitude
B. always approximates that of one degree of latitude
C. approximates that of one degree of latitude at the equator
D. approximates that of one degree of latitude at 45 degrees North Latitude

24. Which one of the following islands is CORRECTLY paired with the continent to which it geographically belongs? 24. D
A. Azores - Africa
B. Cape Verde - Europe
C. Galapagos - Australia
D. Andaman - Asia

25. Which one of the following pairs is INCORRECT? 25. D
A. Quebracho - tanning extract
B. Cinchona bark - quinine
C. Henequen - cord
D. Cacao - cocaine

KEY (CORRECT ANSWERS)

1. A
2. C
3. B
4. C
5. C

6. B
7. C
8. A
9. B
10. B

11. B
12. D
13. C
14. C
15. A

16. B
17. B
18. C
19. B
20. A

21. A
22. A
23. C
24. D
25. D

TEST 2

DIRECTIONS: Each question or incomplete statement is followed by several suggested answers or completions. Select the one that BEST answers the question or completes the statement. *PRINT THE LETTER OF THE CORRECT ANSWER IN THE SPACE AT THE RIGHT.*

1. Of the four major areas of concentration of population, the smallest is to be found in the — 1. B
   A. central and eastern portions of China
   B. northeastern region of the United States
   C. Indian subcontinent
   D. central and southern portions of Europe

2. An arid area that has been made productive by means of irrigation is the — 2. B
   A. Atacama Desert
   B. Imperial Valley
   C. Sahara Desert
   D. Arabian Desert

3. Which one of the following has the largest water-power potential? — 3. B
   A. South America
   B. Africa
   C. Asia
   D. North America

4. Of the following, the country which has the greatest amount of cultivable land per capita is — 4. A
   A. Canada
   B. Australia
   C. United States
   D. Russia

5. Which of the following groups lists countries adversely affected by mercy slumps in world price levels due to major dependence on the export sales of a single commodity? — 5. D
   A. Norway, Sweden, Denmark
   B. Japan, Philippines, Indonesia
   C. Greece, Turkey, Albania
   D. Bolivia, Brazil, Colombia

6. Which one of the following factors necessary for industrialization does Japan lack? — 6. D
   A. Manpower
   B. Power resources
   C. Suitable climate
   D. Iron ore deposits

7. Which one of the following types of climate is characterized by cool, damp summers and mild, wet winters? — 7. C
   A. Tropical rainforest
   B. Mediterranean
   C. Marine west coast
   D. Humid continental

8. There is a significant difference between the North and South Polar regions with respect to — 8. C
   A. temperature
   B. vegetation
   C. altitude
   D. precipitation

9. The outstanding advantage of the Mercator projection is utilized in 9. A
   A. charting nautical courses
   B. showing the distribution of economic resources
   C. giving accurate representation of the overall shape of continents
   D. indicating the earth's major topographical features

10. The outstanding topographic feature of Eurasia is 10. A
    A. a great east-west barrier of mountains, plateaus, basins, and deserts
    B. an immense north-south plain flanked by highlands
    C. an absence of navigable rivers and choice harbors
    D. its location in predominantly tropical and equatorial latitudes

11. In which one of the following groups are the cities closest to the same parallel of latitude? 11. C
    A. New York - Berlin - London - Warsaw
    B. New York - Berlin - Frankfort - Paris
    C. New York - Istanbul - Madrid - Naples
    D. Antwerp - London - Vienna - Warsaw

12. Which one of the following is NOT used by cartographers to represent topographical relief? 12. A
    A. Isobars B. Color C. Hachures D. Contour

13. Within an hour, a given point on the earth passes through 13. A
    A. 15°
    B. 30°
    C. 30° at the Tropic of Cancer
    D. a varying number of degrees in accordance with changes in latitude

14. Through which body of water would an oil tanker pass FIRST en route from Kuwait to France via the Suez Canal? 14. C
    A. Arabian Sea
    B. Indian Ocean
    C. Persian Gulf
    D. Red Sea

15. Which one of the following would be the LAST port of call of a vessel using the St. Lawrence Seaway from the Gulf of St. Lawrence to the iron ore fields at the western tip of Lake Superior? 15. B
    A. Toronto B. Duluth C. Cleveland D. Milwaukee

16. For most of their length, the Tigris and Euphrates Rivers flow through 16. B
    A. Iran
    B. Iraq
    C. Syria
    D. Saudi Arabia

17. Which one of the following ports is normally ice-blocked during the winter? 17. B
    A. Hammerfest, Norway
    B. St. Petersburg, Russia
    C. Murmansk, Russia
    D. Petsamo, Russia

18. Which one of the following has BOTH a larger area and a larger population than the other three? 18. B
A. Argentina B. Brazil C. Mexico D. Chile

19. Which one of the following items describes the major use of land in the United States today? 19. A
A. Pasture and grazing land
B. Forests and woodland, not pastured
C. Cropland
D. Urban areas, parks, highways, and other public facilities

20. Which one of the following statements about the geography of Canada is INCORRECT? 20. B
A. Most of the country lies north of 50° latitude.
B. Canada has a population roughly twice that of New York State.
C. Most of Canada's topography is hostile to permanent settlement.
D. Since World War II, there have been important discoveries of oil, iron ore, and non-ferrous metals.

21. Which one of the following statements about Latin America is INCORRECT? 21. A
A. The rich soils of tropical Latin America offer a major resource for agricultural expansion.
B. In general, the major commercial products of the Latin American countries use only a small percentage of the gainfully employed workers.
C. In the Amazon Basin, a major handicap to economic development is lack of manpower.
D. The high cost of transportation is a serious deterrent to industrialization.

22. The topographical factor that has most impeded exploration of the interior of Africa is the 22. A
A. existence of falls and cataracts on the large rivers
B. presence of a series of immense rift valleys on the east coast
C. tropical rainforest of the Congo region
D. existence of immense stretches of arid and desert lands in Northern Africa

23. The Middle East, with its wealth of oil deposits, 23. A
A. is hampered by the scarcity of cultivable land in the area
B. is self-sufficient with respect to coal and iron
C. is considered potentially capable of industrial development comparable to that of Western Europe
D. has surpassed the United States as the world's leading oil production region

24. The summer monsoon is usually accompanied by heavy rainfall because 24. D
   A. coastal mountain ranges prevent the moist air from rising
   B. the land cools much more rapidly than the surrounding waters
   C. monsoon winds generally are part of the prevailing Westerlies
   D. light, warm, moist air rises as it passes over hot lands

25. Which one of the following statements most accurately describes a major difference between Australia and New Zealand? 25. A
   A. Australia is drought ridden; New Zealand has considerable rainfall.
   B. Australia is densely populated; New Zealand is sparsely populated.
   C. The bulk of Australia's population comes from Western Europe; the bulk of New Zealand's population comes from the Asiatic mainland.
   D. Australia's industries are primarily extractive; New Zealand's economy is geared to grazing.

KEY (CORRECT ANSWERS)

1. B
2. B
3. B
4. A
5. D

6. D
7. C
8. C
9. A
10. A

11. C
12. A
13. A
14. C
15. B

16. B
17. B
18. B
19. A
20. B

21. A
22. A
23. A
24. D
25. A

TEST 3

DIRECTIONS: Each question or incomplete statement is followed by several suggested answers or completions. Select the one that BEST answers the question or completes the statement. *PRINT THE LETTER OF THE CORRECT ANSWER IN THE SPACE AT THE RIGHT.*

1. The climate which possesses the greatest seasonal contrasts in temperature is the — 1. B
   A. Desert
   B. Humid Continental
   C. Tundra
   D. Polar Ice Cap

2. The map projection that is best for calculating directions is the — 2. C
   A. Azimuthal B. Polar C. Mercator D. Gnomonic

3. The savanna climate is distinguished by — 3. C
   A. rainfall of less than 4 inches annually
   B. maximum rainfall during the winter season
   C. annual rainfall of 30-60 inches
   D. no month with average temperature above 32° Fahrenheit

4. Hachures on contour lines indicate — 4. A
   A. direction and degree of slope
   B. atolls
   C. depressions
   D. drumlins

5. The leading factor in the expansion of the chemical industry in the Gulf Coast area of the United States is its — 5. A
   A. great endowment of raw material
   B. favorable climate
   C. location amid great mass markets
   D. great supply of technically trained personnel

6. The leading export crop of the United States is — 6. A
   A. cotton B. corn C. wheat D. tobacco

7. The area that rivals the Pittsburgh district for leadership in production of iron and steel is — 7. B
   A. Birmingham area
   B. South Chicago - Indian Harbor - Gary area
   C. Youngstown - Sharon area
   D. Eastern Pennsylvania

8. The winter wheat belt of the United States is — 8. A
   A. on the eastern slopes of the Great Plains, west of the corn belt, and north of the cotton belt
   B. located in the Northern Great Plains, from western Minnesota to central Montana

C. situated in the area from the middle of Ohio to central Kansas and Nebraska
D. extends from southern Pennsylvania and Virginia west through Missouri

9. The leading export of Canada is 9. C
A. wheat B. non-ferrous metals
C. pulp and paper products D. petroleum products

10. Which one of the following statements about South American trade is CORRECT? 10. B
A. There is a great deal of intra-continental trade.
B. Its main exports are raw materials and foodstuffs.
C. Its export trade is relatively insensitive to world economic conditions.
D. Its commercial relations continue to be chiefly with Spain and Portugal.

11. The most democratic and politically stable republic of Central America is 11. D
A. Honduras B. Guatemala
C. Nicaragua D. Costa Rica

12. South America 12. B
A. extends from 55 degrees south latitude to 12° south latitude
B. has its greatest width through the tropics
C. is partly in the frigid zone
D. has at least 50% of its area in temperature zones

13. A reason for Moscow's industrial leadership in Russia is 13. A
A. its controlling position in the transportation network of Eastern Europe
B. its importance as a seaport
C. its tremendous supply of raw materials
D. its enormous amounts of power

14. The leading coal producing area of the former U.S.S.R. was the 14. D
A. Ural area B. Tula area
C. Kuznetsk area D. Donets basin

15. Which of the following descriptions of European rivers is INCORRECT? 15. C
A. Rhine - excellent navigability; direction is right for trade
B. Rhone - best natural highway through mountains of South Europe
C. Volga - a unifying influence; gives access to high seas
D. Thames - example of a broad estuary forming excellent harbor

16. The Norwegian Coast is ice-free all year round because of 16. B
A. latitude
B. prevailing westerlies
C. great barrier reef
D. deep fiords

17. An important element in the economy of Germany is that it 17. B
A. has sufficient iron ore for its own needs
B. has its greatest mineral wealth in the form of coal
C. is self-sufficient in petroleum
D. is self-sufficient in copper

18. Throughout its history, the economy of Iraq has centered about the _____ Rivers. 18. B
A. Nile and Niger
B. Tigris and Euphrates
C. Ganges and Brahmaputra
D. Mekong and Irrawaddy

19. Which one of the following is a CORRECT statement? 19. B
A. Teheran is in Saudi Arabia.
B. Damascus is in Syria.
C. Beirut is in Jordan.
D. Baghdad is in Lebanon.

20. In the Mediterranean region, cattle raising is handicapped by 20. A
A. inadequate pasturage
B. lack of demand for beef
C. hot climate
D. high cost of labor

21. The largest Moslem state in the world, in terms of population, is 21. D
A. Egypt
B. Libya
C. China
D. Pakistan

22. *It is second ranking in world production of tin and first as an exporter of copra and pepper. It is also known for its production of petroleum and tobacco.* 22. A
This excerpt is BEST applied to
A. Indonesia
B. Malaya
C. Philippines
D. India

23. Which one of the following statements is NOT correct for Myanmar and Thailand? 23. B
A. Each has a great north-south river that dominates the life and economy.
B. Both have comparatively high population densities.
C. Each nation is an important exporter of rice.
D. Both have large seaport cities located on river deltas.

24. Which one of the following rivers is most important as a means of access to the interior of China? 24. C
A. Hwang Ho (Yellow)
B. Si Kiang
C. Yangtze Kiang
D. Amur

25. Which one of the following statements most accurately describes Africa? 25. A
A. Most deficient in plains, consisting essentially of one great plateau
B. Few large rivers, each having little water during a major portion of the year
C. Meager hydroelectric power potential, chiefly available during the rainy months
D. Essentially covered with tropical jungle vegetation with clearings mainly in the desert regions

---

KEY (CORRECT ANSWERS)

1. B
2. C
3. C
4. A
5. A

6. A
7. B
8. A
9. C
10. B

11. D
12. B
13. A
14. D
15. C

16. B
17. B
18. B
19. B
20. A

21. D
22. A
23. B
24. C
25. A

TEST 4

DIRECTIONS: Each question or incomplete statement is followed by several suggested answers or completions. Select the one that BEST answers the question or completes the statement. *PRINT THE LETTER OF THE CORRECT ANSWER IN THE SPACE AT THE RIGHT.*

1. In the United States, standard time belts or zones are spaced at longitude intervals of 1. A
   A. 15° B. 10° C. 30° D. 45°

2. Which of the following statements referring to the Mercator projection is NOT true? 2. C
   A. All the meridians are drawn as straight north-south lines.
   B. Navigators find this projection convenient to plot their courses at sea.
   C. Accurate representation with a minimum of distortion for all land areas is achieved.
   D. Greenland appears as large as Africa.

3. The chief advantage of using the *homolosine* projection maps is that 3. B
   A. directions are indicated better than on Mercator's projection
   B. equal geographical areas are represented by equal map areas
   C. location of places by means of longitude and latitude is more easily determined than on the older maps
   D. the difficulties of representing a spherical surface on a flat piece of paper are obviated

4. Indicate the statement listed below which is CORRECT. 4. D
   A. In the horse latitude belts, the air ascends to higher levels and is thereby cooled.
   B. The equatorial calm belt is a dry belt.
   C. In a cyclone, the pressure of the air becomes increasingly greater from the circumference toward the center, where the pressure is greatest of all.
   D. India's plentiful rains during the summer are caused by the monsoons.

5. Closest to the Antarctic continent is 5. D
   A. Tasmania B. Cape of Good Hope
   C. New Zealand D. Cape Horn

6. The typical crops of a region frequently reflect distinctive features of the environment. Appropriate illustrations of this generalization would include all of the following EXCEPT 6. B
   A. corn in Indiana and Illinois
   B. winter wheat in North and South Dakota

C. olives in the Mediterranean lands
D. rice in monsoon Asia

7. The world's leading producer of nickel is 7. C
A. United States of America
B. Zaire
C. Canada
D. Russia

8. Which of the following leading wheat producing nations ranks lowest in the export of wheat? 8. D
A. Canada
B. United States
C. Argentina
D. Russia

9. Which of these natural products has faced the LEAST competition from recently developed synthetics? 9. D
A. Rubber
B. Quinine
C. Chile saltpeter
D. Tin

10. Which statement is NOT true regarding Italy's geographic handicaps? 10. C
A. It has few navigable rivers.
B. It has a deficiency of coal and iron.
C. It lacks water power potential.
D. The relatively unbroken coastline provides few natural harbors.

11. Indicate the statement below which is NOT true of Spain and Portugal. 11. D
A. Although these countries have coal resources, coal is imported.
B. Exports and imports are similar in nature.
C. The leading industrial cities of these countries are located either on the coast or close to it.
D. Sheep raising is carried on in Portugal but not in Spain.

12. Forests are the most valuable resource of the countries of northern Europe. At the present time, these forests are 12. C
A. not being exploited because of more accessible timber sources
B. being exploited at a rate that promises early exhaustion
C. the basis of extensive industrial development in the region
D. declining in importance as demand for timber lessens

13. Outside Europe, a Mediterranean type of climate is found in 13. B
A. Chile, California, New Zealand
B. Australia, Chile, South Africa
C. Australia, Brazil, California
D. California, Mexico, Argentina

14. Which group of statements about the Mediterranean Basin is CORRECT? 14. C
   I. It has a wide annual range of temperatures.
   II. Volcanic and seismic activity is common.
   III. Its western half is encircled by mountains.
   IV. The Mediterranean type of climate extends well inland.
   V. It is a region of small, sheltered valleys.

   The CORRECT answer is:
   A. I, III, IV
   B. I, II, IV
   C. II, III, V
   E. II, IV, V

15. Which one of the following statements concerning the industrialization of Australia after World War II is MOST correct? 15. D
   A. There has been very little industrialization.
   B. More people are sustained by pastoral pursuits than by manufacturing.
   C. More people are sustained by agricultural pursuits than by manufacturing.
   D. Manufacturing now sustains more people than do agriculture and pastoralism.

16. In which of the following pairs is Canada the leading producer of the world? 16. C
   A. Petroleum and natural gas
   B. Wheat and corn
   C. Asbestos and nickel
   D. Wood pulp and rayon

17. The Saar is 17. A
   A. south of the Ruhr
   B. adjacent to the Ruhr
   C. a region rich in iron ore
   D. the source of the coal used in the Ruhr industries

18. Land utilization in Syria and Israel reflects the fact that 18. C
   A. the mountains are so close to the sea that there is no arable land along the coast
   B. the climate is too dry for cereal crops
   C. the mountains are a source of water for irrigation
   D. grazing is more profitable than agriculture in so dry a climate

19. One of the most important economic ties which link Turkey and the United States today is the fact that Turkey is a major supplier of 19. B
   A. bauxite B. chrome C. platinum D. titanium

20. Because of the distribution of natural resources in Europe, 20. C
   A. France must import bauxite
   B. Germany can export iron ore
   C. Sweden can export iron ore
   D. Denmark can export coal

21. The United States and Russia differ most markedly in that 21. B
   A. Russia lacks important coal reserves; the United States has large coal reserves
   B. the United States has an excellent system of natural waterways; Russia has not
   C. Russia includes an extensive semi-arid region; the United States has no region of similar climate
   D. the United States is a cotton exporting country; Russia imports almost all the cotton it requires

22. Development of an iron and steel industry in Brazil is hampered by the fact that 22. C
   A. Brazil lacks iron ore
   B. Brazil has iron ore but no coal
   C. the coal deposits of Brazil are far from the iron ore deposits
   D. Brazil lacks iron ore and coal

23. In the humid Pampa region of Argentina, 23. A
   A. grazing is being gradually replaced by other types of land use
   B. cultivation of wheat and corn is being gradually replaced by grazing
   C. the hot moist climate has been utilized for coffee production
   D. cotton is rapidly becoming the dominant crop

24. The significant reasons why Sao Paulo is the leading coffee region of Brazil are that 24. D
   I. coffee grows best in the rich loamy soil found in Sao Paulo
   II. the altitude is well suited for coffee production
   III. the wet and dry seasons of the uplands are favorable to coffee growing
   IV. governmental and private capital have encouraged coffee production

   The CORRECT answer is:
   A. I, III
   B. II, III, IV
   C. I, II, IV
   D. I, II, III, IV

25. The island of Bahrein is of significant geographical interest because of its 25. B
   A. strategic position
   B. petroleum resources
   C. famous Moslem shrines
   D. productive date and coffee crops

KEY (CORRECT ANSWERS)

1. A	6. B	11. D	16. C	21. B
2. C	7. C	12. C	17. A	22. C
3. B	8. D	13. B	18. C	23. A
4. D	9. D	14. C	19. B	24. D
5. D	10. C	15. D	20. C	25. B

TEST 5

DIRECTIONS: Each question or incomplete statement is followed by several suggested answers or completions. Select the one that BEST answers the question or completes the statement. *PRINT THE LETTER OF THE CORRECT ANSWER IN THE SPACE AT THE RIGHT.*

1. Which of the following cities is INCORRECTLY matched with the locational advantage responsible for its growth? 1. A
   A. St. Louis - Roads tend to converge near passes to take advantage of breaks or low spots in mountain barriers
   B. St. Paul - Roads tend to converge at the head of navigation on a river
   C. London - Roads will tend to focus at a ford nearest the mouth of a river
   D. Shanghai - Ocean routes focus at or near the mouth of a large river that offers access to the interior

2. A glance at the map reveals that South America has 2. D
   A. the greatest part of its land in a temperate zone
   B. a comparative absence of mountain and plateau areas
   C. a coast line for every country except one
   D. one of the finest potential systems of inland waterways in the world

3. Which of the following was NOT a contributory factor to the decline in importance of the Amazon Valley as a producer of rubber following World War I? 3. A
   A. Disease of the hevea trees
   B. The introduction of the use of synthetic rubber
   C. The competition from the rubber plantations of the Orient
   D. Poor efficiency of the Brazilian workers

4. Which one of the following statements about China is TRUE? 4. C
   A. Her potential cropland in remote areas can be greatly increased.
   B. She has excellent reserves of high grade coal, copper, and iron.
   C. Her regional geographical position is excellent.
   D. Communist agriculture reforms have solved the problem of an inadequate food supply.

5. Which one of the following statements about African geography is FALSE? 5. A
   A. Most of its area south of the Sahara is in dense forest.
   B. There is little intra-continental trade.
   C. A mediterranean type of climate can be found at both its northern and southern ends.
   D. There are desert lands both north and south of the Equator.

6. The map projection that presents the world with the least distortion of the shape and the areas of the land masses is _____ projection. 6. D
   A. the Mercator
   B. Mollweide's homolographic
   C. the equatorial orthographic
   D. Goode's homolosine

7. The International Date Line is nearest 7. D
   A. Greenland B. Madagascar
   C. Iceland D. New Zealand

8. An isobar on a map indicates places having 8. C
   A. equal temperature B. equal rainfall
   C. equal pressure D. the same prevailing winds

9. The Great Circle air route from Seattle to Tokyo passes closest to 9. B
   A. Guam B. the Aleutian Islands
   C. the North Pole D. the Hawaiian Islands

10. The South American countries that do not have an outlet (port) on either the Atlantic or Pacific Oceans are 10. C
   A. Bolivia and Uruguay B. Paraguay and Peru
   C. Bolivia and Paraguay D. Peru and Uruguay

11. The Danube River flows into the 11. B
   A. White Sea B. Black Sea
   C. Caspian Sea D. Bay of Biscay

12. Arrived at Bremen, a freighter carrying a mixed cargo of wheat, merino wool hides, frozen beef, and gold probably shipped direct from 12. D
   A. Yokohama B. Gibraltar C. Colombo D. Sydney

13. The greatest production of uranium in North America occurs in 13. B
   A. New Mexico
   B. the Great Bear Lake Region
   C. the Sudbury Region
   D. the Montana-Wyoming Region

14. The major iron and steel producing area in the southern part of the United States is 14. C
   A. Memphis B. Atlanta
   C. Birmingham D. Savannah

15. The two most important mineral deposits in western Germany are 15. C
   A. coal and petroleum B. magnetite and copper
   C. coal and potash D. magnetite and aluminum

16. Of the following sources of electric power, the one from which the greatest amount of electricity is being produced in the United States at the present time is 16. D
    A. water power B. fuel oil
    C. natural gas D. coal

17. Which of the following areas has the greatest potential of water power for the production of electricity? 17. A
    A. Africa B. Australia
    C. Asia D. South America

18. Copper refineries have been located near New York City chiefly because it has 18. D
    A. a favorable climate
    B. a surplus of highly skilled labor
    C. an abundant supply of cheap power
    D. the largest market for the consumption of copper

19. Which of the following statements regarding the geography of Africa are TRUE? 19. B
    I. The irregular coastline provides numerous harbors.
    II. The northern coast is a broad coastal plain merging into the Sahara Desert.
    III. Johannesburg is the largest city in South Africa.
    IV. Rubber is a product of the Congo area.

    The CORRECT answer is:
    A. II, III B. III, IV
    C. I, II, III, IV D. I, II, IV

20. Despite the complexity of Asian geography, in teaching that continent special emphasis may correctly be placed on the 20. D
    A. pervading problem of summer drought caused by the monsoon
    B. vast lowland areas of central Asia as an untapped agricultural frontier
    C. limited influence of the monsoon as compared with the influence of the prevailing westerlies
    D. rainfall pattern in south and east Asia caused by the monsoon

21. A study of the world's supply of coal would reveal that 21. A
    I. there is a dearth of coal in the Southern Hemisphere
    II. North America ranks first and Asia second in known reserves of coal
    III. present known reserves of coal indicate the need for developing other sources of power within the next hundred years
    IV. anthracite coal is fairly evenly distributed throughout the Northern Hemisphere

    The CORRECT answer is:
    A. I, II B. II, III C. I, IV D. III, IV

22. The steel industry is generally located near coal principally because 22. C
   A. most coal mines are in the vicinity of seaports
   B. the iron needed in the manufacture of steel is usually found in deposits close to coal mines
   C. the bulk of the coal needed to make steel is many times the bulk of iron
   D. labor is cheaper in coal areas than in iron areas

23. The technique most frequently employed in dry farming is 23. B
   A. silting
   B. fallowing
   C. flood-watering
   D. irrigating

24. The diurnal temperature range is greatest in 24. A
   A. tropical deserts
   B. humid continental regions
   C. areas of Mediterranean climate
   D. tropical rainforests

25. Possession by an area of resources of salt, sulphur, lime, and coal makes possible industrial development in 25. B
   A. plastics
   B. chemicals
   C. glassware
   D. metal alloys

KEY (CORRECT ANSWERS)

1. A
2. D
3. A
4. C
5. A
6. D
7. D
8. C
9. B
10. C
11. B
12. D
13. B
14. C
15. C
16. D
17. A
18. D
19. B
20. D
21. A
22. C
23. B
24. A
25. B

ANSWER SHEET

NO. ______________ PART ______ TITLE OF POSITION ____
(AS GIVEN IN EXAMINATION ANNOUNCEMENT - INCLUDE OPTION, IF ANY)

E OF EXAMINATION ____ DATE ______________
(CITY OR TOWN) (STATE)

RATING

USE THE SPECIAL PENCIL. MAKE GLOSSY BLACK MARKS.

1 A B C D E	26 A B C D E	51 A B C D E	76 A B C D E	101 A B C D E
2 A B C D E	27 A B C D E	52 A B C D E	77 A B C D E	102 A B C D E
3 A B C D E	28 A B C D E	53 A B C D E	78 A B C D E	103 A B C D E
4 A B C D E	29 A B C D E	54 A B C D E	79 A B C D E	104 A B C D E
5 A B C D E	30 A B C D E	55 A B C D E	80 A B C D E	105 A B C D E
6 A B C D E	31 A B C D E	56 A B C D E	81 A B C D E	106 A B C D E
7 A B C D E	32 A B C D E	57 A B C D E	82 A B C D E	107 A B C D E
8 A B C D E	33 A B C D E	58 A B C D E	83 A B C D E	108 A B C D E
9 A B C D E	34 A B C D E	59 A B C D E	84 A B C D E	109 A B C D E
10 A B C D E	35 A B C D E	60 A B C D E	85 A B C D E	110 A B C D E

Make only ONE mark for each answer. Additional and stray marks may be counted as mistakes. In making corrections, erase errors COMPLETELY.

11 A B C D E	36 A B C D E	61 A B C D E	86 A B C D E	111 A B C D E
12 A B C D E	37 A B C D E	62 A B C D E	87 A B C D E	112 A B C D E
13 A B C D E	38 A B C D E	63 A B C D E	88 A B C D E	113 A B C D E
14 A B C D E	39 A B C D E	64 A B C D E	89 A B C D E	114 A B C D E
15 A B C D E	40 A B C D E	65 A B C D E	90 A B C D E	115 A B C D E
16 A B C D E	41 A B C D E	66 A B C D E	91 A B C D E	116 A B C D E
17 A B C D E	42 A B C D E	67 A B C D E	92 A B C D E	117 A B C D E
18 A B C D E	43 A B C D E	68 A B C D E	93 A B C D E	118 A B C D E
19 A B C D E	44 A B C D E	69 A B C D E	94 A B C D E	119 A B C D E
20 A B C D E	45 A B C D E	70 A B C D E	95 A B C D E	120 A B C D E
21 A B C D E	46 A B C D E	71 A B C D E	96 A B C D E	121 A B C D E
22 A B C D E	47 A B C D E	72 A B C D E	97 A B C D E	122 A B C D E
23 A B C D E	48 A B C D E	73 A B C D E	98 A B C D E	123 A B C D E
24 A B C D E	49 A B C D E	74 A B C D E	99 A B C D E	124 A B C D E
25 A B C D E	50 A B C D E	75 A B C D E	100 A B C D E	125 A B C D E

ANSWER SHEET

TEST NO. ________ PART ______ TITLE OF POSITION ____________________

(AS GIVEN IN EXAMINATION ANNOUNCEMENT - INCLUDE OPTION, IF ANY)

PLACE OF EXAMINATION ____________________ [redacted] DATE ______

(CITY OR TOWN)

RATING

USE THE SPECIAL PENCIL. MAKE GLOSSY BLACK MARKS.

1 A B C D E	26 A B C D E	51 A B C D E	76 A B C D E	101 A B C D E
2 A B C D E	27 A B C D E	52 A B C D E	77 A B C D E	102 A B C D E
3 A B C D E	28 A B C D E	53 A B C D E	78 A B C D E	103 A B C D E
4 A B C D E	29 A B C D E	54 A B C D E	79 A B C D E	104 A B C D E
5 A B C D E	30 A B C D E	55 A B C D E	80 A B C D E	105 A B C D E
6 A B C D E	31 A B C D E	56 A B C D E	81 A B C D E	106 A B C D E
7 A B C D E	32 A B C D E	57 A B C D E	82 A B C D E	107 A B C D E
8 A B C D E	33 A B C D E	58 A B C D E	83 A B C D E	108 A B C D E
9 A B C D E	34 A B C D E	59 A B C D E	84 A B C D E	109 A B C D E
10 A B C D E	35 A B C D E	60 A B C D E	85 A B C D E	110 A B C D E

Make only ONE mark for each answer. Additional and stray marks may be counted as mistakes. In making corrections, erase errors COMPLETELY.

11 A B C D E	36 A B C D E	61 A B C D E	86 A B C D E	111 A B C D E
12 A B C D E	37 A B C D E	62 A B C D E	87 A B C D E	112 A B C D E
13 A B C D E	38 A B C D E	63 A B C D E	88 A B C D E	113 A B C D E
14 A B C D E	39 A B C D E	64 A B C D E	89 A B C D E	114 A B C D E
15 A B C D E	40 A B C D E	65 A B C D E	90 A B C D E	115 A B C D E
16 A B C D E	41 A B C D E	66 A B C D E	91 A B C D E	116 A B C D E
17 A B C D E	42 A B C D E	67 A B C D E	92 A B C D E	117 A B C D E
18 A B C D E	43 A B C D E	68 A B C D E	93 A B C D E	118 A B C D E
19 A B C D E	44 A B C D E	69 A B C D E	94 A B C D E	119 A B C D E
20 A B C D E	45 A B C D E	70 A B C D E	95 A B C D E	120 A B C D E
21 A B C D E	46 A B C D E	71 A B C D E	96 A B C D E	121 A B C D E
22 A B C D E	47 A B C D E	72 A B C D E	97 A B C D E	122 A B C D E
23 A B C D E	48 A B C D E	73 A B C D E	98 A B C D E	123 A B C D E
24 A B C D E	49 A B C D E	74 A B C D E	99 A B C D E	124 A B C D E
25 A B C D E	50 A B C D E	75 A B C D E	100 A B C D E	125 A B C D E